Disputed Ground

Disputed Ground

*Farm Groups That
Opposed the New Deal
Agricultural Program*

by Jean Choate

McFarland & Company, Inc., Publishers
Jefferson, North Carolina, and London

LIBRARY OF CONGRESS CATALOGUING-IN-PUBLICATION DATA

Choate, Jean.
    Disputed ground : farm groups that opposed the New Deal
Agricultural Program / Jean Choate.
        p.   cm.
    Includes bibliographical references and index.

    ISBN 978-0-7864-1184-9
    softcover : 50# alkaline paper ∞

    1. Agriculture — United States — History — 20th century.
2. Farms, Small — United States — History — 20th century.
3. Agriculture and state — United States — History — 20th
century.   4. Pressure groups — United States — History — 20th
century.   5. New Deal, 1933–1939.   6. United States —
Economic policy — 1933–1945.   I. Title.
HD1761 .C445   2002
338.1'873'09043 — dc21   2002277790

British Library cataloguing data are available

Cover photograph: President Roosevelt greeting a farmer on his
way to Warm Springs, Georgia. (Courtesy of Franklin D. Roosevelt
Library, Hyde Park, N.Y.)

Manufactured in the United States of America

McFarland & Company, Inc., Publishers
  Box 611, Jefferson, North Carolina 28640
  www.mcfarlandpub.com

Contents

Preface

There has been a long history of agricultural unrest in America, beginning with Bacon's Rebellion, Shay's Rebellion, the Whiskey Rebellion, the Granger protests, the Alliances, the Populist Movement, the Nonpartisan movement of North Dakota and surrounding states, and the Farm Holiday Movement, and continuing up to the present day. The stories of such movements have been told in books such as Szatmary's *Shay's Rebellion*, Buck's *The Granger Movement*, Pollack's *The Populist Response to Industrial America*, Morlan's *Political Prairie Fire: The Nonpartisan League, 1915–1922*, Danbom's *The Resisted Revolution*, Hicks' *Agricultural Discontent in the Middle West, 1900–1939*, and others. But one area of agricultural unrest which has not been chronicled is the opposition of some farm groups to the New Deal agricultural programs of the 1930s.

This story deals with seven groups who opposed Franklin D. Roosevelt's agricultural programs. They are the Missouri Farmers Association, the Farmers Union, the Farm Holiday Movement, the Farmers Independence Council, the National Farmers Process Tax Recovery Association, the Corn Belt Liberty League and the Farmers Guild. Many of the farmers who belonged to these groups and their leaders had originally supported Roosevelt in 1932. But when their ideas were neglected, they began to protest.

One group that supported the New Deal agricultural program is the American Farm Bureau. Their story is told in Campbell's *The Farm Bureau and the New Deal, 1933–1940: A Study of the Making of National Farm Policy*.

Historians often say that history is written by the winners. But this is the story of people who struggled and lost. I came across their stories in letters to organizers of relatively unknown farm organizations— the National Farmers Process Tax Recovery Association, the Corn Belt Liberty League, the Farmers Guild — as well as letters to more familiar groups,

1

such as the Farmers Union, the Farm Holiday Association and the Missouri Farmers Association.

Along the way, I received help from many people. Among those who freely offered assistance were archivists and librarians at the University of Colorado, the Universities of North Dakota and South Dakota, Wisconsin State Historical Society, University of Iowa, Kansas State University, University of Missouri, the Franklin D. Roosevelt Library and the Carl Albert Congressional Archives. It was Stanley Yates, Chief Archivist of Iowa State University Library, who first collected some of the National Farmers Process Tax Recovery Association letters and encouraged me to use them.

I owe a debt of gratitude to my professors at Iowa State University who were especially helpful: George McJimsey, Andrejs Plakans, Dorothy Schwieder and Alan Marcus.

I appreciate the encouragement and helpful suggestions given by Iowa farmer/rancher John Puderbaugh.

I need to thank the Everet Dirksen Foundation, Iowa State Historical Association, Northern Michigan University, Franklin and Eleanor Roosevelt Foundation, and the Carl Albert Center of Congressional Archives for grants that enabled me to travel to various sites to pursue my studies. I appreciate the South Dakota Historical Society, which publishes the *South Dakota History* journal, for allowing me to use a chapter that was originally published in the summer 1992 issue of *South Dakota History*. I also wish to thank the administration and faculty at Coastal Georgia Community College for their helpful support.

Above all, I am thankful for the patience and encouragement of my husband, Woodrow, and children: Anne, Michael, Ruth and Susan.

A Word About Sources

Since the sources are already cited in the endnotes for each chapter I will not attempt here to duplicate that list. Rather, I will describe some of the book's major sources, which can be divided into oral histories, manuscripts, periodicals, books, and photographs.

When possible, I interviewed people who had participated in the various farm movements. One of the most interesting sources was Edward E. Kennedy, who had participated in the Farmers Union as national secretary, the National Farmers Process Tax Recovery Association as lobbyist, and the Farmers Guild as founder. I spent several hours with him in his home in Laurel, Maryland, and taped the interview. The tapes from that interview are in the archives at Iowa State University. I also interviewed the children of A.J. Johnson, the secretary of the National Farmers Process Tax Recovery Association. A.J. Johnson's children continue to farm on the old farmstead in western Iowa. The tapes for this interview, too, are stored at Iowa State. Tapes made by earlier historians are at the Wisconsin State Historical Association Library and at the University of Iowa archives. These were used, but to a lesser extent.

Manuscript collections dealing with farm organizations are available in a number of university archives. My search began at the Iowa State University, and I still recall my excitement upon discovering 39 boxes of letters and reports dealing with the National Farmers Process Tax Recovery Association in the archives there. Some of the boxes seem a little dull, with page after page of farmers' reports on the number of hogs they sold and the number of acres of corn they grew, but others command the attention with letters that vividly describe the plight of the farmers in these difficult years.

At the University of Iowa archives, the Milo Reno correspondence has been put on microfilm, a great help to anyone dealing with his massive correspondence during the heyday and decline of the Farm Holiday

Association. At the University of North Dakota, there are many manuscripts dealing with William Lemke, and at the University of Missouri there is a collection of materials on William Hirth in the 1930s. There may be materials on Hirth at the Missouri Farm Association offices in Columbia, Missouri, as well, but the officials there seemed unwilling to open them up.

The archives at Kansas State University hold the Casement papers, which are the best sources for information on the Farmers Independent Council. For the National Farmers Union, there are great amounts of letters, records, and publicity materials stored at the archives of the University of Colorado. The Carl Albert Center for Congressional Studies and the Western History Collection also hold valuable material; both are located at the University of Oklahoma. The Western History Collection holds much material on John Simpson, the president of the Farmers Union.

Our National Archives hold vast quantities of materials deposited by the Department of Agriculture, and it is wise to secure the assistance of archivists there before attempting research. Another rich source of information is the Franklin D. Roosevelt Library at Hyde Park.

The periodicals I used are varied. There are farm journals, organizational newsletters, newspapers, both local and national, and finally some historical journals. I used the *Aberdeen Evening News, Agricultural History, the American Liberty Magazine, Annals of Iowa, Dakota Farmer, Des Moines Register, Farmers Union Herald, Iowa Union Farmer, the Mississippi Valley Historical Review, the Missouri Farmer, the National Union Farmer, the New York Times, the Oklahoma Union Farmer, Social Justice and South Dakota History Journal.*

The books I used include Christiana McFayden Campbell's *The American Farm Bureau and the New Deal, 1933–1940* (Urbana: Illinois University Press, 1962); Ray Derr's *Missouri Farmers in Action* (Columbia: Missouri Farmers Press, 1958); Lowell K. Dyson's *Red Harvest: The Communist Party and American Farmers* (Lincoln: University of Nebraska Press, 1982); Gilbert Fite's *George N. Peek and the Fight for Farm Parity* (Norman: University of Oklahoma Press, 1954); Richard Kirkendall's *Social Scientists and Farm Politics in the Age of Roosevelt* (Columbia: University of Missouri, 1966); Dale Kramer's *The Wild Jackasses: The American Farmer in Revolt* (New York: Hastings House, 1956); Theodore Saloutos' *The American Farmer and the New Deal* (Ames: Iowa State University Press, 1982); Saloutos and John D. Hicks' *Twentieth Century Populism: Agricultural Discontent in the Middle West, 1900–1939* (Lincoln: University of Nebraska Press, 1951); John Shover's *Cornbelt Rebellion: The Farmers Holiday Movement* (Urbana: University of Illinois Press, 1965); and

John A. Simpson's *The Militant Voice of Agriculture* (Oklahoma City, Oklahoma, Mrs. John A. Simpson Co., 1934).

The photographs are all from the Farm Security Administration and are housed in collections at the Library of Congress, the Franklin D. Roosevelt Library at Hyde Park, and the South Dakota State Historical Society. The hundreds of farm photographs in the collections make a very interesting visual presentation of the history of the period. Many of them are available for viewing online. Making a choice among the hundreds of wonderful photographs was difficult.

Prologue

The fields were green, fresh with the promise of spring—green pastures, plowed fields with new plantings of corn, wheat, or beans coming out of the ground. Each mile along dirt roads, there were white country farmhouses, white or red barns and corncribs, chicken coops, pig pens. And all were set within fences to keep cattle in. These settled homesteads were all part of a dream, the dream of farmer pioneers who settled the land two, perhaps three generations before.

The Native Americans who lived here before had held other dreams—of woods and meadows, hunting grounds, movable villages and small plots of corn, beans, squash and pumpkins. But they had been forced to move off of these lands, and their dreams had been replaced by the dreams of the new settlers, the American farmers.

But the dreams and labor of the farmers had not been realized. Something was wrong with their Eden. The prices they received for their produce were not sufficient to enable them to live well, despite their long hours of toil. In their disappointment, they turned to farm organizations and farm leaders who held out promises of a better life.

Life on the farm had been hard in the 1910s and 1920s but became even harder in the 1930s. With the election of Franklin D. Roosevelt to the presidency, and the agricultural programs of the New Deal, a struggle developed between the government and some of the leaders of the agricultural organizations. It was a struggle for the farmer's allegiance and a questioning of the motives and methods on both sides. It was a struggle to determine the future of American agriculture. It was, in truth, disputed ground.

1

William Hirth and the Missouri Farmers Association

One early twentieth century farm organization was the Missouri Farmers Association. It was formed by William Hirth, a Missouri farm editor who was deeply concerned about the problems of the farmers of his region.

As a young boy Hirth came with his parents to live on a farm 1½ miles from the small town of Rush Hill in Autrain County, north central Missouri. He grew up on that farm knowing the day to day work of chores, plowing, planting, hoeing, cultivating, haying and corn picking.[1] As he was fond of saying, he knew farm life, he had "tasted ... its joys and sorrows and its failures and triumphs."[2]

In Rush Hill, Hirth came into contact with the Farmers' Alliance. He joined a local unit and observed the group as it struggled with ways to help the farmer.[3]

One of the ways was to participate in cooperative buying ventures.[4] Members of his county association ordered flour, machine oil, binding twine and harvesting machinery through the Alliance.[5]

Hirth also imbibed the populist philosophy of the Alliance, reading "everything bearing on the farmer's troubles" that he could find. He argued and debated in the local meetings, and at eighteen was elected lecturer for the Rush Hill union.[6] In 1896 he campaigned for William Jennings Bryan, stumping the state for him.

In later years, Hirth's writings would continue to repeat many of the things which the farm leaders of the Populist Revolt of 1896 had said. Like Mary Ellen Lease, the Populist leader who told the farmers to "raise less corn and more hell," Hirth constantly urged the farmers of Missouri to organize to achieve fairer prices for their products. In later years he would call it the principle of "production cost and a reasonable profit."[7]

9

Following three years in college, a few years selling life insurance and some time spent studying law, Hirth and a partner bought a newspaper, *The Columbia Statesman*. They operated the *Statesman* for nearly five years. During this time Hirth devoted considerable attention in the paper's columns to agricultural issues; he suggested the establishment of a creamery in Columbia, and supported farmers' institutes. He also continued to write articles about William Jennings Bryan, who remained his idol.[8]

Hirth then sold *The Statesman* and brought out a new paper called *The Missouri Farmer and Breeder*. In his first issue of the paper in October of 1908 Hirth stated that his purpose was to "Reduce the drudgery of farm life, to make two dollars grow where one dollar grew before and to spread the glories of good old Missouri to the four winds."[9]

In 1912 the name of the paper was shortened to simply *The Missouri Farmer*. The early articles in *The Missouri Farmer* were largely articles about better farming methods. But after a few years Hirth began to include articles that dealt with farm prices, which he felt were too low. And he began to write about the advantage of farmers joining together cooperatively. He included articles about cooperative societies in Wisconsin, Minnesota, Nebraska and adjoining states.[10]

In an article in *The Missouri Farmer*, February 1914, he wrote: "Farmers are at a disadvantage because of their independent action, the lack of organization and cooperation. The individual farmer may raise his voice against an injustice of railroad rates, stockyard charges, or the exorbitant prices of the Standard Oil Trust, but what does it amount to? But let a million, or ten million ... protest ... through united organization and something will be done."[11]

Hirth suggested that farmers in Missouri should organize small farm clubs. A farmer in Chariton County, Arron Bactel, who subscribed to *The Missouri Farmer*, decided to follow Hirth's suggestions. He wrote to Hirth and asked him to send some copies of the magazine in which he had written his latest proposals. Hirth sent the papers. Bachtel distributed them among his neighbors and asked them to meet at the Newcomer School House the next week. They met on a cold March evening in the small one-room schoolhouse and discussed the possibility of forming a club, but reached no decision. The next week, March 10, 1914, seven men agreed to band themselves together into a farm club. The group grew slowly, but in the fall they agreed to put in a combined order for binder twine, which they placed through Hirth. Hirth also ordered coal from a West Virginia Coal Company. The price of coal went up, but the farmers in the farm club obtained coal for the old price. When news of this got around, farm clubs began to spring up in other counties in Missouri.[12]

Hirth, at his own expense, sent out material to people who were interested in forming groups, suggesting club constitutions and by-laws and a speech to be read. He traveled around the state, urging farmers to join together in what was called the farm club movement.[13]

He also continued to write about the growth of the farm club movement in *The Missouri Farmer*. In his writings Hirth told how the members of the farm clubs could pool their purchasing power by placing orders for things he could contract for with suppliers, such as binder twine, salt, feed, coal, machinery, etc.[14]

There were many farmers in Missouri who had been members of other older farm groups, such as the Farmers' Alliance or the Grange, who were familiar with the idea of farmers working together cooperatively. So Hirth was not preaching an entirely new idea to them. In fact, some of the older groups still existed in scattered sections of the state and needed only to be reinvigorated with the enthusiasm of the farm club movement.[15]

At this time the economic status of farmers was reasonably comfortable. Increased demand for farm products from the nation's growing urban population led to rising prices for agricultural goods. There was also a growing market for American farm products in Europe.[16] Farmers turned to the farm clubs not out of desperation but because the idea offered educational, social and financial benefits.

By 1915 Hirth began to discuss the possibility of linking the various individual farm clubs in county organizations and eventually a state-wide organization.

Following his recommendations, the first county-wide meeting of farm club members was held in the County Court Room in Warrenton, Missouri. Attendance was good, and farmers crowded in the room, "sitting in windows, standing in aisles," and some who were not able to get into the room stood in the outside corridors.

Speaking at this first county meeting, Hirth told his audience that he expected farm clubs to meet twice a month and provide definite subjects for discussion at each meeting. He would send materials to the clubs to help them with their programs. Hirth also shared with this first county meeting his hope that the group would grow so that it could "market the products of the farm direct to the consumer and thus throw off the stranglehold which the packers and others now maintain upon the throat of producer and consumer alike."[17]

The group elected a farmer living near Warrenton as president of the County Association. He secured carload prices on bran, oats, hay, corn and other feeds. The county farmers joined together to order these things because they could order in bulk at lower prices.[18]

In September of 1916 Lafayette County Farmers also formed a county association.[19] The farm club movement continued to spread throughout the state of Missouri. By 1916 there were farm clubs organized in twelve or more counties. In his paper Hirth wrote: "Every mail brings letters ... telling about Farm Clubs which have been formed or are in the process of being formed."[20] By the beginning of 1917 there were at least 500 clubs organized throughout the state.[21]

To Hirth the farm clubs were important not only because they saved farmers money and helped them sell their goods at higher prices, but also because they strengthened rural life in other ways. Hirth suggested that lively discussions at farm club meetings would "relieve monotony" and bring "more enjoyment into farm life."[22] He also printed with appreciation a letter from a Missouri farmer who wrote that the farm club had brought "a friendlier, more neighborly spirit" among the farmers in his community.[23]

In January of 1917 a statewide Missouri Farmers Association was formed. Nearly 500 farmers attended the meeting and discussed the projects that they wished to see included in the new organization. Officers were elected. Generally, the first presidents of the group were farmers who were active leaders in their local and county associations.

In an article in his magazine Hirth wrote that this new organization was "designed to put a Farm Club into every school house of the state." At the beginning it would mean supplying seeds, fertilizer, corn, hay, potatoes, coal, salt, etc. to members at low prices, which alone might save millions of dollars to Missouri farmers. But he believed the real purpose of the organization was not only to help farmers buy things at low prices but also to "so band ourselves together that we shall have something to say about what the fruits of our sweat and toil shall bring in the market place."[24]

In keeping with Hirth's philosophy, and probably at his urging, the group issued various resolutions. One resolution urged farmers to rouse themselves "from lethargy and indifference." Another resolution called for the building of elevators and the formation of livestock shipping organizations. The formation of legislative committees to pressure the state legislature concerning agricultural issues was suggested. *The Missouri Farmer* was accepted as the official organ of the new association.[25]

Soon the county associations developed livestock shipping organizations. Farmers drove their livestock to a nearby town where small stockyard holding pens had been built. There the combined stock of several farmers would be shipped out to the stockyards.[26]

Much of this activity is reminiscent of cooperative efforts of the

Grange or the Farmers' Alliance, but in one respect the Missouri Farmers Association was different. There were to be no farm club stores. The farmer members were to purchase only a limited number of items through their clubs. Their main purpose was to work toward achieving better prices for the goods they sold cooperatively through their organizations.

This was a major emphasis of Hirth's. In *The Missouri Farmer* he warned: "The 'Farmers Stores' of the ... old days all failed and I don't want us to go into the same ditch." In the same vein he wrote: "In the first place, years ago, the Grange, Farmers Alliance and 'Wheelers' went into the ditch through the establishment of Farmers' Stores."[27]

In articles in *The Missouri Farmer* Hirth listed the reasons why the old Farmers' Stores had failed. They had failed because the farmers had hired inexperienced managers, because the farmers had invaded a field in which they were unfamiliar, and because farmers failed to patronize their own stores when price cutting started from competing stores in the community.[28]

The Missouri Farmers Association early showed an interest in politics. In 1918 a group from Missouri, which included William Hirth, went to Washington to speak before the Agricultural Committee in the Senate. They went to ask for relief from what they considered inequities in the cattle and hog feeder program of the Food Administration.

In 1919 the Missouri Farmers Association hosted a dinner for the members of both houses of the Missouri state legislature. Following the dinner the group presented the state senators and representatives with a list of five demands for legislative action. Among their demands were a "fair and adequate cooperative law."

The agricultural cooperative law, which the Association had sponsored, was passed by the 1919 legislature.

By 1920 the Missouri Farmers Association had purchased or built 100 local elevators and 200 warehouses. One hundred fifty Missouri Farmers Association livestock shipping associations had been organized and were operating. A cooperatively owned flour mill was expected to be completed in the next year. Hirth also predicted that the Association would build several poultry and egg processing plants and locate them in small towns across the state.[29]

From the very beginning of the Farm Club movement, Hirth had expressed his concern that farmers were not receiving an equal share of American prosperity. Even in the prosperous war years Hirth felt that this was true. In 1919 he wrote an editorial entitled "Just Suppose" in which he suggested that farmers should go on strike and demand the wages that street car conductors and motormen had recently won in wage bargaining in Chicago. Why, he said, if farmers were to make that kind of wage for their time and get a five percent return on their investment in land and tools, farmers should be selling wheat for "something like $5.00 a bushel and hogs and cattle around 30c per lb. on the hoof." The prices which he quoted were far above the prices farmers were receiving for their goods at this time.[30]

In 1919 Hirth was also concerned because he foresaw that the demand for American farm products was likely to decrease. He said that the European nations had gone into debt during the war, and he predicted they would be forced, following the war, to sell as much as possible and buy as little as possible. This would be a blow to the American farmers who had become accustomed to selling millions of dollars of their surplus beef, pork, flour, etc., to European markets. Furthermore, American farmers also were facing the competition of cheap beef coming into the United States from South America. He said that the time would soon come when in sheer self defense farmers would begin to demand an "American Market for the American Farmer."[31]

In 1920 American farmers saw the prices of their goods decline disastrously. The price of corn per bushel declined from $1.59 in 1919 to $ .62 in 1920, then to $.55 in 1921. It rose to $.73 in 1922, to $.88 in 1924, and then fell again to $.75 in 1925.[32]

Hog prices per hundredweight fell from $12.42 in Middle Western markets in 1920 to $7.13 in 1921, $8.06 in 1922 and $6.53 in 1923.[33]

The decrease in crop prices led to a decline in the value of agricultural property. In Missouri the estimated total value of farm land in 1920 was three million dollars, in 1925 the estimated value of farmland had fallen to two million dollars.[34]

While farmers saw the prices of the goods they sold decline, and the value of their farms decrease, the prices of everything else seemed to rise.

These prices had a serious impact on farmers everywhere. Bankruptcies among farmers increased from 6% of the total number of farmers in 1920 to 9% in 1921, 14% in 1922 and 17% in 1923.[35] Hirth wrote a state representative that in 1926 there were approximately 28,500 vacant farm homes in Missouri, and "unless relief comes speedily the same fate will overtake thousands of other farms."[36]

Farmers and farm leaders across the country looked for some ways to relieve the situation. Many said that they wanted to achieve "the cost of production plus a profit."[37] This had been the old rallying cry of the Missouri Farmers Association and the Farmers Union.

George N. Peek, former president of a farm implement company, and others proposed the McNary-Haugen bills, which called for government determined prices for farm products sold within the country and a dumping of the surplus overseas. The first McNary-Haugen bill was introduced into the House of Representatives and failed to pass there, although it had been endorsed by more than two hundred farm organizations.[38]

Following the defeat of the first McNary-Haugen bill in 1924, supporters of the bill decided to call a meeting to discuss the next steps. Farm leaders who met to work out the program for the meeting included Peek, Henry A. Wallace and William Hirth. Representatives of all the major farm organizations attended the meeting, and they voted to form a new farm organization to be called the "American Council of Agriculture." They said that they meant to "secure the enactment of ... legislation embodying the principles of the McNary-Haugen bill and thus secure for American agriculture equality with industry and labor."[39]

Writing to Representative Clarence Cannon of Missouri that agriculture was "dying," Hirth warned that unless the farmers across the country were contented and prosperous, they might turn from supporting the nation's constitution to striving to destroy "our venerable institutions."[40]

A year later, when a revised McNary-Haugen bill was not reported out of committee in the House in time to be passed, Milo Reno of the Iowa Farmers Union suggested a meeting of farm leaders in Des Moines, Iowa. His idea was endorsed by the National Farmers Union.[41]

Thirty seven charter members met and suggested the formation of the Corn Belt Committee. The three resolutions Of the Committee were: 1) Construction of the farmers' own marketing machinery, including grain co-ops and terminal elevators; 2) Cost of production for the farmers' crops; 3) Creation of an export corporation to 'buy up' agricultural surpluses.[42] These were very similar to the proposals embodied in the McNary-Haugen bills.

William Hirth was an early member of the Corn Belt Committee and supported its programs with enthusiasm. In an article in *The Missouri Farmer* Hirth wrote, "I believe that in the years to come this group of men will become more or less historic, for it was through them that the 24 ranking farm organizations of the Corn Belt agreed to 'bury the hatchet' and to henceforth fight their battles side by side."[43] For several years, Hirth was chairman of the group.[44]

In addition to chairing the Corn Belt Committee, Hirth attended numerous sessions of agricultural committees in Washington in his efforts to secure the passage of the McNary-Haugen bills.[45]

In a letter to John Napier Dyer of the National Horticultural Council, Hirth stated, "I have submitted more defense on behalf of the various farm relief bills than any other half dozen farm leaders in the Country, and

I am profoundly convinced that genuine equality for Agriculture cannot be accomplished except through a measure like the McNary-Haugen Bill."[46]

While Hirth was attending various conferences and meetings of House and Senate agricultural committees, a new figure was emerging as a leader in the Missouri Farmers Association. Howard Cowden, as Secretary of the Association, conducted much of the day to day business of the Missouri Farmers Association.[47]

Cowden apparently appointed many of the field workers who supported him to the executive committee of the Missouri Farmers Association, and when differences between Hirth and Cowden developed, the executive committee had supported Cowden.[48]

In 1926 the Missouri Farmers Association at their annual convention passed a resolution that field men who were presently serving on the executive committee could not be reappointed, and no more field men were to be appointed to the board.[49]

Cowden submitted his resignation in March of 1927 to become effective at the annual Missouri Farmers Association Convention in September of 1927. However, a majority of the executive board still supported him, and the matter of who was in control of the Association was not resolved until the 1928 convention. There, pro-Cowden supporters were booed, and Hirth was elected President of the Association by an overwhelming majority.[50]

Other events also occupied Hirth's attention in 1928 and 1929. He was interested in starting cooperative oil and gasoline stations, and began work on organizing such stations.[51]

In politics, Hirth, having worked hard for the passage of the McNary-Haugen bills, was angered by President Coolidge's vetoes of the bills. Hirth felt that Coolidge had been influenced by Hoover, who was serving as Secretary of Commerce at the time, to write the vetoes. Therefore, he was especially angry when Hoover secured the Republican party's nomination for president in 1928.[52]

Hirth devoted several pages of *The Missouri Farmer* to criticizing the Republican Party for failing to live up to the promises they had made to the farmers in earlier platforms. He said they were offering to the voters the same promises, "undoubtedly upon the theory that the Republican party has handed the empty plate to the farmer so often they can probably take a chance and pass it once more." Hirth supported the candidacy of Al Smith for president.[53]

Following the election of Hoover, Hirth barely gave him time to get started in office before he began to criticize his farm legislation. In April of 1929 he wrote to Senator Norris: "I can surmise your disappointment, not to say disgust, over the so-called farm relief bill which has been introduced in the House, and which supposedly reflects the wishes of the White House."[54]

Hirth continued to actively oppose the Hoover administration, writing in *The Missouri Farmer* that he believed the Federal Farm Board created by Hoover was "seeking to destroy the old farm organizations in an effort to clear the way for the re-election of President Hoover."[55]

He also told his readers in *The Missouri Farmer* that few "city people realize the true economic position of Agriculture," and that it was a "mere statement of fact to say that except for our coal miners, farmers, as a class, are enduring a lower standard of living than any other great class in the Nation, and as long as this is true, a cultural rural existence is out of the question."[56]

Hirth felt that the current situation of farmers in Missouri and the rest of the nation demanded new policies.

In 1931 Hirth, in an editorial in *The Missouri Farmer*, wrote that some leaders were urging farmers to "keep out of politics." However, he felt that it was "politics that has brought Agriculture to the profound crisis which now confronts it, and only politics will get our ox out of the ditch."[57]

Hirth was operating from a sense of desperation because of the effects of the Depression on the farmers in Missouri. Writing to Charles Dawes, president of the Reconstruction Finance Corporation, he commented, "I want to say to you in confidence that in my opinion the situation out in the rural districts is constantly becoming more dangerous. With hogs selling for approximately 3 cents per lb. on the farm and No. 2 eggs at 6 cents or 7 cents per dozen, we are approaching mighty close to a complete smash."[58]

Hirth visited Roosevelt in New York and was much impressed by him. Upon his return to Missouri, Hirth wrote Judge Caverno: "Incidentally I took supper with Governor and Mrs. Roosevelt night before last and I hope you are for him, for I think he is the best bet in sight, and his heart is absolutely right."[59]

In March of 1932 Hirth, in *The Missouri Farmer*, strongly urged the presidential nomination of Franklin Roosevelt by the Democratic Party. He said that Roosevelt was the "first choice of the overwhelming majority of sincere Democratic leaders ... and likewise strong with farmers as thousands of letters to Democratic members of Congress indicate."[60]

Although Missouri's Democratic delegation was already pledged to a favorite son candidate, Hirth worked to persuade the Missouri Democratic delegation to vote for Roosevelt as a second choice. Hirth planned the printing of numerous articles, mailing these materials out to delegates and acting in other ways to secure Missouri's shift on the second ballot.[61]

Franklin D. Roosevelt, campaigning by train, addresses a small-town audience. (Courtesy of Franklin D. Roosevelt Library, Hyde Park, N.Y.)

Following the nomination of Roosevelt, Hirth called it a "victory for the 'forgotten man.'" It was a victory against the special interests who had worked to defeat Roosevelt's nomination because of his concern for the "distressed millions on the farm" and the "distressed millions in the cities."[62]

Later that year Hirth made speeches in support of Roosevelt throughout the Middle West, taking a five day junket through several states and also making five radio speeches. Hirth told Roosevelt that he was "shelling the Corn Belt with radio addresses."[63]

It is likely that Hirth's stand influenced the Missouri Farmers Association when they adopted their convention resolutions that year. They met in August of 1932, a time when many farmers in Iowa, Minnesota, South Dakota and other middle western states were planning a Farm Holiday strike. But they did not agree to participate in the strike. Instead, they issued a resolution which stated, "While we deeply sympathize with the farmers who are involved in the so-called 'farmers strike' ... we do not believe that this effort can terminate successfully." Rather, they recommended that the farmers make known their dissatisfaction with their situation at "the ballot box and not on the public highways."[64]

On election day Hirth wrote Roosevelt a long letter discussing the problems which the farmers faced and suggesting that a solution to the farm problem would be the establishment of the cost of production which had so long been a rallying cry of the Missouri Farmers Association as well as the Farmers Union. Hirth stated, "It is my firm conviction that the safety of our Nation in the years to come demands not only that we must assure a wholesome and secure existence to the millions who now live upon the land, but that the land must offer an ever increasing welcome to those whom the machines eliminate in the cities, for where else can the latter find refuge?"[65]

Roosevelt answered the letter on January 17, 1933: "Your letter ... was buried in the flood of congratulatory messages.... I know you will understand and excuse the delay."[66]

Probably Hirth envisioned for himself a prominent role in the newly elected president's program for agriculture. He was concerned, however, as he began to recognize that Roosevelt was discussing agricultural programs with university professors from eastern colleges. Hirth wrote to a friend that the president is "surrounding himself with professors and hairsplitters who are long on theories, but who don't know what it means to have a bloody head as I have from daily contact with closed banks and the marketing of millions of dollars of farm commodities."[67]

Although Hirth wanted to influence Roosevelt's agricultural program, he said he did not want to become Secretary of Agriculture. One of his leading supporters in Congress, Representative Clarence Cannon from Missouri, promoted his candidacy. But Hirth wrote to Roosevelt: "I understand that some of my Missouri friends have written you suggesting me for Secretary of Agriculture and I want you to know that I have not directly or indirectly inspired this suggestion — I think a President should be permitted to select his Cabinet free from outside wire pulling."[68]

There is some debate as to whether Hirth really wanted to be Secretary of Agriculture and was just playing what he thought was the polite game. He very probably had mixed emotions. The recent struggles within the Missouri Farmers Association may have made him fear that leaving the organization to go to Washington would endanger his status in the Association.

Whether he actually wanted the position of Secretary of Agriculture or not, Hirth hoped to have some influence on farm policy. But he unwisely refused to participate in planning meetings before the election, replying to W.R. Ronald, who invited him to such a meeting in Chicago, that he felt planning should wait until after the election, and "if [Roosevelt] wins then we should get our heads together and strive to perfect a practical plan that will assure the farmers of the country an adequate price in our home markets."

The phrasing of his reply suggests that Hirth was hoping Roosevelt would propose a farm program similar to the McNary-Haugen proposals of the 1920s. He did try to make suggestions concerning farm policy to Roosevelt's campaign advisors, Louis Howe and James Farley. Perhaps he hoped they could convince Roosevelt to back his ideas for agriculture.[69]

Later, he was disappointed and very angry when he was not invited to participate in a national conference to plan for the New Deal farm program. On December 12 and 13 a conference was held in Washington to draw up the farm program for the New Deal. The conference was chaired by Henry Morgenthau, Roosevelt's valued friend and advisor.[70]

Many agricultural leaders were invited, including Edward O'Neal of the Farm Bureau, John Simpson of the Farmers Union, Henry A. Wallace, and Rexford Tugwell, university professor. But Hirth, although he had frequently stated that he wanted to be included in such deliberations, was not invited.

Not only was he angry because he was not invited, Hirth also did not approve of the program which the conference adopted. Most of the members of the group approved the domestic allotment program, in which farmers would be paid to produce smaller amounts of crops. But Hirth still clung to his ideas of tariff protection and marketing agreements as promulgated in the McNary-Haugen bills.

In a letter to Representative Cannon, Hirth wrote that he would not even accept the office of Secretary of Agriculture if it were offered because of the new proposals for agriculture. "I would spurn it if it should be offered me under the half-baked plans for Agriculture which now seem to be looming on the horizon."[71]

Nor did Hirth approve of Roosevelt's nominee for Secretary of Agriculture, Henry A. Wallace of Iowa. Hirth said Wallace was a "hair splitter, and I think such a man at the head of Agriculture at this time would be nothing short of tragic." He felt that the head of the Department of Agriculture should be a man who "has hair on his breast."[72]

As details of the proposed New Deal farm legislation, the Agricultural Adjustment Act, became known, William Hirth began to attack it in editorials in The Missouri Farmer. In an April 1, 1933, editorial Hirth published an open letter to Senator Bennett Champ Clark. He said that the current plans were an "unsound farm proposal." He opposed the reduction of current surpluses because many people in the United States were hungry. He also believed that if foreign tariff walls were broken down, the surpluses could be used in trade with Europe. Neither did he approve of those parts of the proposal which he felt would result in the employment of a "horde of inspectors."

According to Hirth, those who supported the New Deal farm program were "for the most part … the same group of 'Yes' men who fronted for the Agricultural Marketing Act, and for Mr. Hoover's ill-fated farm board." On the other hand, those who had strongly supported the McNary-Haugen bills, such as Ricker, Talbott, Murphy, Keeney, Ward, Reno or Hirth, had not even been invited to the conference that planned the New Deal's farm program.[73]

Articles in *The Missouri Farmer* in the 1930s continued to point out the problems of agriculture. In one 1933 article, entitled "Farm Wages Downward," the writer stated that the average wage paid per month to a farm worker with board was $16, as opposed to $21 a year before.[74]

In the Women's Department, a section devoted to news of the W.P.F.A. (the women's auxiliary of the Missouri Farmers Association), several articles describe the aprons, curtains, rugs, etc. which could be made from flour bags.[75]

Hirth said the New Deal farm programs were failures. In various editorials and features in *The Missouri Farmer*, he pointed out problems in the New Deal's agricultural program. He also made fun of the program as much as possible. In May of 1934 he published a poem entitled "Birth Control on the Farm," a paraphrase of James Whitcomb Riley's "Little Orphan Annie," with an oft-repeated line, "The allotment plan will get ya if ya don't watch out."[76]

In the same issue he published a cartoon which had originally been published in *The New York Herald Tribune*. It was entitled "Pursuing Agricultural Adjustment to its Logical Conclusion." Consisting of four panels, the first showed Secretary Wallace with an AAA axe heading for a long line of baby pigs. In the second panel Wallace is pulling a cow off to the stockyards. In the third panel he is talking to the farmer while they survey the ruined farm. Wallace says, "Ah, I have an inspiration!" And in the fourth panel he takes off after the farmer with his axe. And the farmer is running for his life.[77]

At the heart of this disagreement between Hirth and the New Deal is a difference in vision of what the rural world should become. Hirth feared that the federal government would take control of agriculture and that only the most successful farmers would be allowed to remain on the farm.

This was very different from Hirth's vision of rural life. He felt that in Missouri and throughout the nation there were great numbers of poor farmers who were doing the best they could. They paid their taxes, raised their children, served in the army when called. "And just because these farmers do not rotate their crops as they really should, and do not do other things in a 'scientific' way as they should, if to them their little farms are 'the sweetest place on Earth,' will we tolerate a time when those who consider themselves the Lord's anointed shall be permitted to say to them, 'Move on —find some other way to make a living.'"[78]

Working despite these fears, Hirth did all he could to strengthen rural life in Missouri. His farm paper, *The Missouri Farmer*, reached 50,000 subscribers. He built up an organization with 18,000 or more members enrolled. He organized cooperatives for selling farmers' products at higher prices and providing farmers' needs at lower prices. He worked to educate the farmers in scientific farming principles. He sought to strengthen their political impact both on the state and national levels. The farm cooperatives which he founded still operate in Missouri today.

In an editorial in an early *Missouri Farmer* he wrote to encourage farmers, saying, "Let's do something!" and exhorting them to blaze "the way for a greater American Agriculture and for a farm home which shall be free and untrammeled in the years to come."[79]

Inspired by his youth on the farm and his early Farmers Alliance dreams, Hirth was a great battler in the fight for a better tomorrow for Missouri farmers, but in his later years he became discouraged as he observed the effects of the Depression on Missouri farms and feared the effects of the New Deal agricultural programs.

2

John Simpson and the Farmers Union

By 1932 many farmers' incomes had dwindled to about half the level that they had been the year before. Milk was selling at 1 cent a pound, hogs at 3 cents a pound, cattle for 5 cents a pound and corn for 10 cents a bushel.[1]

In 1931 Oklahoma Senator Elmer Thomas sent out letters to many of his constituents requesting information concerning farm conditions. Many farmers replied, and their letters tell of their desperate situations. One farmer wrote in November of 1931: "A large percentage of farms here have been sold for taxes ... men are losing their farms to the loan companies and no relief in sight."[2]

Another wrote: "If I didn't get my living from our cows and hens we just wouldn't live, if we depended on our farm. Last year the crop about payed the taxes and a small upkeep. You can see that a farm has been a job without an income."[3]

Still another wrote: "I will say that the agricultural situation here is very bad, for all the farmers are in debt and no way to pay up."[4] Some farmers, in addition to describing their situation, voiced a desperate threat. One farmer wrote: "Please do something for us or tell them we will do it ourselves."[5]

As the farmers in the 1930s attempted to deal with the Depression, various farm leaders responded with plans for improving their lot. Franklin Roosevelt and the Department of Agriculture developed a program which was also intended to improve agricultural conditions. Between these two there was often a struggle for farmers' allegiance, and a questioning of motives and methods. It was disputed ground.

As the Oklahoma farmers wrote to Thomas in 1931 concerning the conditions in their area, they often mentioned an Oklahoma farm leader whose ideas they trusted, John Simpson. Simpson had been born on a farm in Nebraska, attended country school, taught for some time and then studied law at the University of Kansas. He received his law degree in 1896 and then married Millie Berlet that same year.

In the turbulent years of the 1890s he was caught up in the activities of the Populist Party, and in 1897 was appointed accountant in the state auditors office in Lincoln, Nebraska.

In 1901 he decided to move his family to the newly opening Oklahoma Territory and drew a claim at the El Reno drawing for land in Caddo County, Oklahoma. He farmed the land, worked as a country banker and served in the Oklahoma State Legislature. During the legislative session he began to realize that individuals had little influence on lawmakers unless they were organized.[6]

On returning home he joined the Oklahoma Farmers Educational and Cooperative Union, often called the Farmers Union, and set out to increase its membership and influence. Under his leadership the Oklahoma Farmers Union grew from 231 in 1917 to over 23,000 in 1921.

During the 1920s he worked tirelessly for farming causes both on the state and national level. He was particularly interested in the McNary-Haugen bill, which would have guaranteed farmers a good price for their products sold in the country and dumped the surplus abroad. When this was vetoed by Coolidge, Simpson worked for the presidential campaign of Al Smith.[7]

In 1930 he became national president of the Farmers Union. The Farmers Union was one of the three leading farm organizations in the United States. The group was founded in 1902 in Raines County, Texas, by a former member of the Farmers Alliance and the Populist Party, Newt Gresham. The group had spread from Texas throughout the Middle West and into many western states. But ever since the middle 1920s there had been a struggle for power in the Farmers Union between two different groups.

Belonging to the first group were the heads of the large cooperatives, who wanted government aid to keep up commodity prices. This group included the Kansas Farmers Union and leaders of the Northwest Organizing Committee. This three-man committee had been formed in 1927 to organize for the Farmers Union in states such as Nebraska, North and South Dakota, and Minnesota.[8] The members of the group were: A.W. Ricker, Saint Paul newspaper editor; C.C. Talbott, North Dakota farmer; and M.W. Thatcher, head of the Farmers Union Terminal Association.

In the second group were those who wanted the government to set prices on agricultural products, prices that would return the farmers cost of production plus a profit.[9] "Cost of production," a favorite term among some members of the Farmers Union, was defined as all the costs of producing a product, such as seed, fertilizer, machinery and labor. It was believed that these costs could be ascertained by government economists and made the official price for major products such as corn, wheat, cotton, rice, etc. This group included John Simpson and his Oklahoma Farmers Union, and Milo Reno and the Iowa Farmers Union. Simpson and Reno were both dynamic leaders and speakers.[10]

In 1928 Charles Huff of Kansas was chosen by the convention. Huff was popular with the elements in the Farmers Union who favored the development of cooperatives.[11] He was a supporter of the National Farm Board, which had been established under legislation promoted by President Herbert Hoover.

But by 1930 the situation appeared so grave that many members of the Farmers Union felt that cooperative associations were not the answer; instead, they wanted the federal government to guarantee the farmers "cost of production." And they turned to John Simpson to lead them.

When Simpson, with the support of Reno, defeated Huff in 1930 to become president of the national Farmers Union it was a revolt based, at least in part, on the farmers' frustration with the problems of the Depression. The prices they were receiving for their products had continued to decline. And many believed that the Farm Board had been of no assistance to them in this crisis.

One farmer, who was secretary of a Farmers Union local group, wrote to Senator Thomas: "Dear Senator, in ans. to your letter will say that conditions are very bad in Harmon Co. And our attitude toward the Farm Board is not at all friendly. Please get rid of them as quick as possible."[12]

Another farmer wrote: "I sincerely can say that the farmers of this district strongly resent the actions of the farm board."[13] Still another wrote: "I am for the Farmers Union plan of a set price on all farm crops. How could the Railroads run their roads if their rate for services should vary each day as prices on farm products do?"[14]

The editors of the *Nebraska Union Farmer* described the election of Simpson to the presidency of the National Farmers Union as a "contest that involved the attitude of the [Union] toward the agricultural marketing act and the Federal Farm Board."[15]

Following his election, Simpson, who had been critical of the Farm Board before his election, stepped up his examination of the Farm Board's operations.

What was the Farm Board? Herbert Hoover, following his election as president in 1928, began to push for legislation that would solve the problems of the farmers by creating a "giant instrumentality clothed with sufficient authority and resources ... [that] would at once transfer the agricultural question from the field of politics into the realm of economics and ... result in constructive action...."[16]

After considerable debate in Congress, during which congressmen from agricultural states tried but failed to add features of the McNary-Haugen plan to the president's bill, the Agricultural Marketing Act was passed. It created a Farm Board, and President Hoover appointed Alexander Legge, the president of the International Harvester Corporation, to head it. Other members of the Farm Board were leaders in various cooperative associations throughout most of the nation.[17]

Congress also appropriated $500 million for the use of the Farm Board. The board could use the money to make loans to cooperative associations. They in turn would buy products from farmers at times when the prices tended to go down, thus raising the general level of prices, and store the produce and sell it for higher prices when prices had risen.[18]

Because of the deepening crisis of the Depression, the Agricultural Marketing Act did not work. Prices kept falling. Alexander Legge, in a speech before the Association of Land Grant Colleges and Universities in 1930, said: "During the past twelve months, what might be termed emergency cases have occupied far too much of our time, to the detriment of the efforts we had hoped to put in on the educational or long-time program."[19]

Charles E. Huff became president of the Farmers National Grain Corporation and secretary of the Grain Stabilization Corporation. The Farmers National Grain Corporation was a cooperative owned by producers. The Grain Stabilization Corporation was an incorporated device of government. At one time the Grain Stabilization Corporation owned 257 million bushels of wheat.[20]

As Huff recounts the events of the times, he says that he believes the Agricultural Marketing Act "was not a strong device, and its best provisions were used sparingly."[21]

There is evidence that some producers responded to the efforts of the Farm Board by increasing their production. In the area of wheat control, there were problems of not enough storage capacity. The reaction of the Farm Board was to urge voluntary reduction.[22]

This excited considerable opposition and finger-pointing among various sections of the nation. According to the writers of the *St. Paul Pioneer Press*, "To attempt to make the spring wheat growers stand the full burden of reduction for the whole industry, to penalize it for the action of the winter wheat branch, would be both unjust and futile, and would arouse a well-grounded resentment among the producers of the Northwest."[23]

In Kansas, Legge was quoted as telling a group of Kansans, "The biggest hog will always lie in the trough. Kansas is now in the trough."[24]

Disagreements over the Farm Board and its operations divided the National Farmers Union. States such as North Dakota, Montana and Minnesota supported the Farm Board, while Nebraska, Iowa and Oklahoma bitterly opposed it. Nebraskans charged that the Farm Board would result in "a great bureaucracy," with control vested in "a political board, answering not to farmers, but to the exploitative masters of the government."[25]

In his annual report to the Farmers Union convention in 1931 Simpson roundly criticized the Farm Board: "No more complete failure of a department of government was ever recorded in the 150 years of existence of this nation than the failure of the Farm Board. They took charge of the marketing of wheat in the Fall of 1929. And the price went down and the general trend has been downward ever since.... They deliberately used, through their setups, taxpayers' money in an effort to control the farm organizations. No farm organization could borrow money from the Board without bowing the head and bending the knee."[26]

In a much publicized incident, Simpson charged that the Farm Board had admitted to a congressional committee that they had the power to raise or lower the price of wheat and had decided to lower it. This was based on information Simpson had received from Senator Thomas. Simpson had asked Thomas to attend a conference between the Farm Board and the Senate Agricultural Committee.

Thomas said that the members of the Farm Board did not really want to make public some of their activities. However, he got a statement from Alexander Legge, Chairman of the Federal Farm Board, in which he admitted that the Board had control of the American wheat situation and that it could "place the price of wheat at any figure which it desired." Legge also stated that when the Board saw the price of wheat going up it had "stepped in and sold 3,500,00, bushels on the Chicago Exchange, which caused a break in the market." Thomas added that the Board told the committee they did not wish to see the price of wheat going up because they wanted to keep it low so that it could be "within the reach of the buying public during these depressed times," and also they desired to keep wheat prices low so that it could be fed to hogs and livestock instead of corn.[27]

When these things were reported in the national newspapers, Legge responded to the charges by calling Simpson a liar. Associated Press articles quoted Legge as saying, "You can tell Mr. Simpson, or any other higher authority that he sometimes quotes as a source of his information, that any man making that statement, whether in public office or out, is just an unmitigated liar."[28]

To which Simpson replied that in his long "experience with life and men" he had found that "he who shouts 'liar' when an accusation is leveled at him usually has little but the power of his voice to clear himself." He went on to say that he believed he could beat Legge in a name calling contest, for while Legge may have gained his experience in yelling at his subordinates at the harvesters trust, he, Simpson, had "driven mules and called hogs to the hog trough." However, Simpson suggested that the best way to solve the issue would be to call for a congressional investigation of the Board's activities.[29]

At about the same time, Milo Reno wrote to Senator Charles McNary, Chairman of the Senate Agricultural Committee, saying that Legge's remarks were "an insult to organized farmers, in general, and the Farmers' Union in particular...." Reno also asked for a congressional investigation of the Farm Board in order to determine the truth of the matter.[30]

On January 3, 1931, Senator McNary made public a letter from Legge stating that it would be possible to raise the price of wheat to $1 a bushel but that "in our judgment this would not be advisable."[31]

In an article in the Oklahoma Union Farmer the editor charged that the Farm Board's admission showed that the Board was "using the distress of the wheat farmers to help the hog and stock growers and also to help the millers and bakers."[32]

In evaluating the activities of the Farm Board, Theodore Saloutos, noted agricultural historian, comments, "The problems the Federal Farm Board faced were insurmountable. Apart from the opposition of the private marketing agencies and the feuding among rival organizations and the larger cooperative associations, it had to contend with the worst depression on record."[33]

James C. Stone, of the Federal Farm Board, issued a statement in May of 1931: "The Farm Board is doing its best to work out the many problems that confront it, and if the people are not satisfied with its work let them change the personnel. I know many of the members would not be sorry to be relieved."[34]

In the spring of 1931, Simpson and his wife traveled in Europe for two months studying farm conditions in Italy, France and Denmark, and attending the World Wheat Conference at Rome. Following his trip, at various times in his speeches he would refer to his travels and make comparisons between the conditions of American farmers and European farmers. He said that those who would exploit American farmers "feed him the bunk about the terrible conditions of the peasants of France. They ask him how he would like to wear wooden shoes as they do in Holland and Denmark."

Simpson said that the poor, debt-ridden American farmers believe this bunk and it makes them feel better. However, according to Simpson, thirty five per cent of French farmers lived in modern homes with electricity and running water, while in the United State only ten per cent of the farmers lived in modern homes.[35]

The European trip reinforced Simpson's ideas about the best ways to deal with agricultural problems. Following his visit to Europe, he fought even harder for government programs to help American farmers.

Simpson continued to speak out concerning the problems the farmers were facing. In July 1931 he gave an address entitled "What's the Matter? What's the Cause? What's the Remedy? over NBC. In the address he stated that something was certainly wrong when "six million heads of families with twenty-four million wives and children depending on them" are out of work and had been for a year.

"Something is the matter when millions of farm women carried eggs to market this very day and took 10¢ per dozen for them, sold their butterfat for not more than 18¢. Something is the matter when thousands of farmers are hauling to market good wheat that would make good wheat bread to feed thirty million hungry people and are receiving less than 40¢ per bushel for it. Something is the matter when a wheat farmer sells a bushel of wheat in Montana for 30¢ bushel and can only take home with him for the 30¢ three of the sixty-five loaves of bread that bushel of wheat will make. Something is the matter when a million and a half farmers have lost their homes by foreclosure since May 1, 1920."[36]

In December of 1931 Simpson testified before the Senate Committee on Agriculture. During his testimony he held up a check for 75 cents received by W.B. Estes for the sale of seven lambs. After receiving his check, Estes had contacted the Department of Agriculture to find out what happened to his lambs. The Department reported that they were sold to consumers for $83.70.

Simpson continued: "On a Pullman dining car coming here from Chicago, I was charged 85 cents for two lamb chops—10 cents more than Estes received for seven lambs.... If you wish to know why discontent is widespread and deep-seated in the farming region, Estes can give you the reason."[37]

Following the publication, T.E. Howard, of the Colorado Farmers Union, wrote to say that a man from the stockyards had called threatening to sue the paper for printing the story.[38]

Simpson replied that Howard did not need to be afraid of any libel suits. He had the seventy-five cent check and the original sales sheet. He also believed that he could prove that the seven sheep might bring as much as $300 when the consumer paid for meat, woolen clothes, etc. Furthermore, papers all over the United States had published the story; so they would have a lot of defendants if they brought suit.[39]

In Texas and Oklahoma, farmers were facing dust storms which began in January of 1932 and brought dust and destruction that spring, restricting travel and laying some sandy farm fields bare. These storms would increase in intensity each year through much of the 1930s.[40] But bad as the weather conditions were, farmers in 1932 were even more concerned with the depressed farm prices that their products were receiving.

In March of 1932 Simpson received a letter from Franklin Roosevelt, then Governor of New York. In his letter FDR said that he wished he could meet Simpson personally and talk with him on matters which he, Roosevelt, "needed information and advice at first hand."[41]

Simpson met with FDR in early April and then wrote him a letter stating that he was glad to hear him say that he would be willing for a plank to be placed in the National Democratic platform "promising the farmers such legislation as would secure to them cost of production for that part of their crops used in this country." He also was pleased that the president said he was in favor of a plank asking for an international conference on the silver question. One of the techniques which Simpson felt should be used to deal with the Depression was inflation of currency, which he felt might be effected by increased coinage of silver. This was an idea strongly supported by the Populists in the late 19th century, and Simpson never quite forgot his Populist roots.[42]

In the early summer of 1932 Simpson began to campaign for Roosevelt. He sent him a copy of an address he made over NBC, and FDR replied, "I appreciate very much the kind words you used concerning me in your speech and I sincerely hope that you have the opportunity and the strength to make many more."[43]

Simpson campaigned wholeheartedly for Roosevelt. In September he wrote Roosevelt that he had spoken at thirty-four meetings over the last four weeks. These included Farmers Union picnics, Holiday Strike meetings and even a Labor Day meeting. The events had been well attended, with only two having less than a thousand persons present and several having more than two thousand. He had also made radio speeches in Iowa, Nebraska and South Dakota. "It just looks like this mid-west country is going to give you a big majority."

But he had a concern: "I have a letter from one of the farm leaders in which he says that the Wilson Allotment Plan is being presented to you." Simpson did not believe the Allotment Plan was what the farmers needed. "The Farmers Union, the Grange, and the Farm Bureau all opposed the Allotment Plan in the last session of Congress." He believed that the plan was formulated by the Chamber of Commerce and that big business was financing Wilson and others to lobby for it in Congress.

He suggested that FDR in his speech on agriculture "stick to the plank in the national Democratic Platform which pledges farmers everything will be done under the constitution to secure them cost of production for their products."

Simpson went on to say that in his talks he compared the Democratic platform and the Republican platform. He told people that the Democratic platform endorsed the principles of the Farmers Union marketing plan and the Frazier refinancing plan, while the Republican platform only promised more Farm Board.

And then he closed his letter by telling of his plans for the next two months. "My meetings continue up to election day and will take me, after this week, into Iowa, North Dakota, Montana, Colorado, and ... Oklahoma."[44]

Concerned about achieving "cost of production" for American farmers, Simpson tried to make sure that FDR was planning to support it if elected. In a conference with Farmers Union officials held in Roosevelt's Pullman Car in Sioux City, Iowa, on September 29, FDR was recorded by Simpson as saying that agriculture was more important to the life of the country than railroads, banks or any other interest. He promised to devote more time to saving agriculture than to saving any other industry. He agreed to see that farmers were refinanced at lower rates of interest and on long-time payment of principal. And he pledged to stand on the Democratic platform plan that promised "everything will be done possible ... to see that farmers get cost of production for their products."[45]

On October 4 A. Berle replied to Simpson's September 6 letter. He said that he thought Roosevelt's speech at Topeka had already stated FDR's position on most of Simpson's questions. Then he continued: "Of course, we all hope that various farm organizations can agree on a program so that a united farm representation can back whatever measure is ultimately proposed."[46]

Simpson wrote Berle that among organized farmers the Topeka speech was "fairly well understood." But he did not think it was so well understood by unorganized farmers. He suggested that the Governor should make a statement publicly endorsing the two agriculture planks of the Democratic platform. "A plain, frank statement will please millions of farmers."[47]

Although they might be considering other farm policies than those which Simpson promoted, Democratic leaders still wanted Simpson's help in securing the farmers' vote. On October 17 James Farley wrote Simpson, "I want you to know that we appreciate all that you are doing and are going to do through the remainder of the campaign."[48] On October 29 Roosevelt wrote to Simpson saying that he had received a "mountain of correspondence," and if letters were any indication, "it will be a sweeping victory in November." He concluded the letter by thanking Simpson for all he had done and assuring him of his warm personal regards.[49]

On December 17, 1932, Simpson wrote President-elect Roosevelt, thanking him for sending Henry Morganthau to represent him at the conference held in Washington by leaders of farm organizations. Simpson said that Morganthau definitely made it known that they should not build a marketing program around the Farm Board. "This policy harmonizes with the Farmers' Union one hundred per cent." Simpson continued to say that, after Morganthau left, an attempt was made to have the conference endorse the activities of the Farm Board. And Simpson said that he, Simpson, had made "successful, emphatic, objection."[50]

Even before the December conference, however, there was talk concerning the possibility of Simpson receiving the appointment of Secretary of Agriculture. As one example of this widespread speculation, he received several requests from publishers for short biographies and statements of his positions.[51]

A few scattered endorsements of Simpson were sent to Roosevelt in November. They were from George Jewett; Aldrich Blake, President of the American Minerals Corporation; North Dakota Senator William Lemke; R.L. Rickerd, President of the Oklahoma Farm Holiday Association; C.C. Talbott, President of the North Dakota Farmers Union; Albert Fickler; the Farmers Union Livestock Commission; and C.H. Hyde, Vice President of the Oklahoma Farmers Union.

By December Simpson was writing close friends, asking for their help in securing endorsements for the position. He felt pretty good about those he had received already. He commented to C.H. Hyde, "I am sure I have more backing up to date than any other person mentioned."[52]

Roosevelt's secretary, Guernsey Cross, wrote Hyde that "Governor Roosevelt is not as yet considering any appointments, but when he does your letter in behalf of John A. Simpson will be before him."[53]

Elmer Thomas also worked to secure endorsements for Simpson. He got a formal endorsement signed by both Oklahoma Senators, the Democratic members of Congress from Oklahoma, the former Governor of Maryland and Representative-elect Will Rogers. In describing his activities,

Thomas commented, "I do not know what chance he has but we are going to see that nothing is left undone to secure the place for him."[54]

However, Simpson was receiving some letters from people saying that they could not endorse him. One such letter came from Edward O'Neal, president of the American Farm Bureau. "We have had many applications for endorsement and our policy is to present the whole matter to our Board of Directors."[55]

Richard Kirkendall, in his book *Social Scientists and Farm Politics in the Age of Roosevelt,* writes that the Farm Bureau was promoting the candidacy of Henry A. Wallace.[56]

One of Simpson's supporters wrote to Hyde that he had "sounded out" Henry A. Wallace. "I got here this P.M. and saw Henry Wallace. He does not seem to be a candidate himself. He is not doing anything either way I guess. Says he will leave that to others. He said Simpson was a fine fellow but did not seem to feel as if he was the right man for Sec. of Ag. He said Simpson was not careful enough about his statements, etc. In other words, he is a little too radical for Wallace, as I took it."[57]

This suggestion that he might be "too radical for the position" was mentioned by others. One friend wrote his congressman "I wish to say for Mr. Simpson that I have known him for many years, and we who know him have only admiration for his sterling honesty and for his thorough knowledge of the Farm problems that confront the farmer today. In some of his ideas he may be a radical, in others he may be a conservative, but whatever he may do, we may depend that he is thoroughly honest in his convictions and will fight to the bitter end for the right as he may see it."[58]

Simpson wrote to a friend, "I have no idea whether or not Governor Roosevelt is giving serious consideration to the requests of my friends. In fact, I would consider it very unusual for a man with as radical views as I have to be placed in a President's cabinet."[59] In another letter Simpson wrote: "Some of the big interests have protested against me as being a radical."[60]

As the rumors circulated that Henry A. Wallace was being considered for the position of Secretary of Agriculture, C.H Hyde wrote to Bernard Baruch to protest. He wrote, "You and I know he [Wallace] is a Republican. You and I know that he supported the Farm Board and their policies until very, very recently."[61]

Resolutions of support continued to pour in. Some were official resolutions, such as the resolutions from the legislatures of Oklahoma, North Dakota and Nebraska. Others came from state Farmers Union organizations as well as small local groups. A typical example is the letter from the Wright County Farmers Union: "If Mr. Simpson is appointed Secretary of

Agriculture it will meet with the satisfaction of a vast majority of the farmers of the United States."[62]

In February W.B. Derk wrote that he hoped God and the President-elect would "give us Men ... Men with strong wills, warm hearts, clear and honest minds and ways. John Simpson, President of the Farmers Union, is one of this sort."[63]

As Simpson corresponded with friends in Oklahoma, he again was hearing tales of the destruction wrought by a year of drought exacerbated by high winds and drifting dust storms. In 1932 there had been a few dust storms. By April of 1933 there would be as many as 179 reported by weather bureau stations.[64]

In late January it was rumored that Roosevelt had asked Henry Wallace, Rex Tugwell and M.L. Wilson to draw up plans for reorganizing the Department of Agriculture. Some people began to guess that was an indication that Wallace was being seriously considered for the post of Secretary of Agriculture. Then, early in February, Roosevelt contacted Wallace, offering him the post. Following this, Raymond Moley called Wallace to repeat the invitation and get his answer. Wallace accepted.[65]

One can only imagine Simpson's disappointment when he learned that, despite his efforts in support of Roosevelt's election and the many endorsements he had collected from farmers, farm groups, and politicians in farm states, he had been passed over for appointment.

More disappointment and disillusion were to come. Soon after the announcement of Henry Wallace as Secretary of Agriculture, a meeting was held to discuss the long anticipated agriculture bill. Simpson did not attend the meeting because he had been "informed at the White House that Congress would recess in a few days for a period of, at least, three weeks and that no agricultural legislation would be taken up before reconvening." Relying on that advice, Simpson had left to fulfill a series of speaking engagements. Learning of the meeting that was held when he was gone, he felt that he had been deliberately deceived.[66]

Much later, Roosevelt wrote Simpson, saying that he hoped Simpson would realize that he had "acted in entirely good faith in telling you that probably Congress would be in session only three or four days."[67]

Following his return, Simpson then wrote the members of the Senate Agricultural Committee asking for the opportunity to come before the committee to present the position of the Farmers Union.[68]

In March 1933 John Simpson spoke before the Senate Agriculture and Forestry committee as it was considering Wallace's New Deal farm legislation. He said that he was President of the National Farmers Union. He also represented the Farmers National Holiday Association. Committee chairman Ellison D. Smith of South Carolina said, "That is universal is it not?" Simpson replied, "It [the Farm Holiday] has more members than any other farm organization that ever existed."[69]

At the hearings Simpson made suggestions to the committee concerning the bill. He hoped to bring the proposed legislation more in line with the cost of production ideas of the Farmers Union and the Farm Holiday. He suggested that the Secretary of Agriculture be empowered to pay the market price for cotton produced in the South. He also opposed the acreage limitation in the program. He said it was impossible to regulate production by regulating acreage. "It is impossible. You would have to have God on your side to be sure that it would work...."[70]

The Senate Agricultural Committee incorporated a cost of production guarantee as recommended by Simpson in the farm bill. But Secretary Wallace was against the cost of production provision. Wallace was saying it would be too hard to operate and that it would be impossible to figure out the cost of production for the various farm products. Farmers asked, "Can not our experiment stations figure the average cost of production?"[71]

In May Simpson wrote Roosevelt. He said that the Senate Agricultural Committee had agreed to include cost of production in the agriculture bill. Later the Senate voted in favor of the bill, including the cost of production provisions. When the bill went to the House, the House refused to concur regarding this particular part of the bill, and thus it went to a conference committee.

Secretary Wallace filed a protest against the cost of production provisions, saying it was economically unsound and also impossible to arrive at such a cost on any definite basis.

Simpson wrote to Roosevelt reminding him that the Democratic platform of 1932 had promised farmers that "everything possible would be done to secure to them cost of production for their products." Simpson stated: "Upon the basis of the promise in that plank I went before the farmers of the Nation in one hundred and eight meetings in twelve states; many times broadcasting over nation-wide hook-ups."[72] In the final outcome, the Agricultural Adjustment Act was passed without cost of production provisions. Excited about the possibility of finally achieving one of the goals of the Farmers Union, Simpson had fought hard, but he had lost.

On May 19 Simpson wrote to Thomas Cashman, "I had the opposition whipped until the President sent word to the House of Representatives that he wanted cost of production eliminated. The leaders in the House told me that unless the President did that they were going to leave it in." He concluded, "That is just one battle in the war.... We may lose this one, but that is no sign we are going to lose the war."[73]

He continued the fight. In July Simpson wrote Roosevelt that a decision had been made barring Farmers' Union institutions from the privilege of borrowing, even though other cooperatives were permitted to do so. He felt that this decision was similar to that made by the Farm Board and had been made by the same attorney. "We feel that in this respect we are not getting a new deal."[74]

As agricultural conditions worsened, on August 16 Simpson sent a telegraph to President Roosevelt:

DEAR MR PRESIDENT CONDITIONS DEPLORABLE IMPROVEMENT LARGELY IMAGINARY STOP THERE WILL BE MORE FARMERS OBJECTS OF CHARITY THIS WINTER THAN EVER BEFORE IN HISTORY OF OUR COUNTRY....[75]

He also suggested to other Farmers Union leaders that they send similar letters.

In September Simpson wrote a long letter to the President stating that he had held 92 meetings throughout the East and Middle West. Each meeting had attracted two or three thousand people. "Everybody connected with agriculture now realizes that what was passed in this part of the Bill [refinancing] does not help a distressed farmer. Foreclosures continue, and the army of homeless farmers is still increasing."[76]

Simpson sent copies of his letter to all the members of Congress, to the Farmers Union papers, and to a number of other papers. He received a form letter from Louis Howe: "Your letter ... [to the President] has been received.... He has asked me to thank you for writing."[77] Simpson also received a letter from Representative Lundeen suggesting that they work together organizing a new political party. Simpson wrote back, agreeing that "the time is about ripe for organizing a new political party." Both seemed to think the name "Farmer-Labor" should be used.[78]

On October 24 Simpson sent a letter to Roosevelt: "'Your Brain Trust' goes from one folly to another. They evidently do not speak the farmers' language, neither do they understand it. The farmers ask for cost of production: and your 'Brain Trust,' through the newspapers, indicate they are going to lend them more money."[79]

FDR responded, "Somehow I had it in my head that you were not given to the kind of extravagant and exaggerated language which never pays either in public or in private life. That is why I was a little surprised by your letter of October twenty-fourth. Perhaps, sometime later on when you are in Washington you will run in and see me and talk things over quietly."[80]

On November 11 Simpson wrote, "Just a little over a year ago, Mr. Herbert Hoover was out here in the West. He was in Des Moines, Iowa. He was well pleased with the reception he received. He could not discover a single person disloyal to him. The poor fellow did not realize the cheering part of his audiences were composed of jobholders. He found it out on election day." Simpson concluded his letter by saying, "I am just trying to keep the wolves off your back."[81]

In November a reporter for *The Christian Science Monitor* asked Simpson what he thought of the Farm Holiday movement led by Milo Reno. Simpson wired back, "Farmers have a right to be mad. It looks like the only way to let people know they have been viciously discriminated against is to break up the furniture and sling the dishes. It is my opinion farmers should refuse and cease to sell their products below cost of production."[82]

Also in November Simpson submitted his report to the National Farmers Union convention held in Omaha, Nebraska. He reminded the group of the historic demands of the Farmers Union concerning taxes, farm boards, military training, etc. He said that he believed in this program with "all his heart and soul." "As your head man, believing in your program, I have worked in season and out of season. I have worked night and day, educatively and legislatively, for the complete fulfillment of those demands that you have made."[83] He was re-elected president of the Union.

In November of 1933 dust storms originating in the southern plains carried dust all the way to Georgia and New York.[84]

About this time Simpson received a letter from the President inviting him to visit him at the White House. In his letter FDR comments that he hopes he has been getting erroneous reports on Simpson's activities and "will be glad to give you the opportunity on your return to bring me first hand information."[85]

Simpson replied that he expected to be back in Washington on the 23rd of December and would call FDR's office and be glad to make him a visit. He continued his letter by telling of the poor prices farmers were receiving for their products: "One man told me he sold three cowhides and lacked ten cents having enough to buy a dog collar."[86]

Simpson continued to criticize the New Deal farm program. He appeared before the conference of mid-western governors in Iowa in March of 1934. Then on March 14, a day after his return from the conference, he was again in action, lobbying in Washington. He went first that morning to the office of William Lemke where he made phone calls setting up appointments for later that day. He chatted briefly with Lemke. As Simpson picked up his papers and briefcase to leave, Lemke asked him to stay. But Simpson replied that he was to appear that morning before the Senate

Finance Committee to give important testimony on issues dealing with farm problems.

After his testimony he walked down the corridors of the Senate Office Building. There he met two friends, Brinkman and Gray of the Grange and the Farm Bureau. They talked for a while and then Simpson said he felt some pain in his chest. They sat down on the top steps. In a few minutes he lapsed into unconsciousness. A nurse was called, and his wife and daughter who lived in an apartment nearby were summoned. An ambulance arrived and he was taken to the hospital. He regained consciousness and talked briefly with his wife and his daughter, who served as his secretary. He died about 4:00 the following morning.[87]

Letters and telegrams were received from public officials, leaders of farm organizations and ordinary farmers throughout the country. From Franklin Roosevelt came a note saying, "Greatly shocked at news of John's death. As you know, he and I have been friends for a long time." From Floyd Olson, Governor of Minnesota: "It was a distinct shock to me to learn of Mr. Simpson's death; particularly in view of the fact that I had seen him just a few days before at Des Moines." From Will Rogers: "American farmers have lost their greatest champion." William Langer, Governor of North Dakota, ordered the state flags to be hung at half mast. Mrs. Morris Self of Ohio wrote, "He has worked and fought for us with all his might and strength. We know he gave his life for our cause." From Rev. D.N. Kelly of the Eureka Farmers Union: "The Moses for the Farmers has gone." Henry Wallace wrote, "Mr. Simpson was a strong figure in the field of those who have given their energies to farm betterment. He very much impressed his personality on us all." From M.F. Dickinson, president of the Arkansas Farmers Union: "A great man is gone."[88]

John Simpson, farmer and farm organization leader, had seen the problems farmers were facing and given all his energy and abilities to try to improve their lot. Because Simpson was trusted by many farmers, his support was solicited by FDR, who hoped that he would convince the usually Republican farmers to support the Democratic ticket in 1932. But after the election Simpson was not chosen to become Secretary of Agriculture. And while FDR's administration recognized that farmers had problems, Simpsons' solutions, such as "cost of production" or increased coinage of silver, were seen as too radical.

Instead, the Roosevelt Department of Agriculture, with the consent of the Farm Bureau and the Grange, proposed the Agricultural Adjustment Act, which sought to limit the production of agricultural staples and reimburse the farmers for limiting their crops. Simpson fought these measures and criticized them when he felt they were ineffective.

John Frederick of Grant County, North Dakota, observes his drought-stricken crops. He expected 20 bushels of wheat from his 40-acre field. (Courtesy of Library of Congress, Washington, D.C.)

And in the summer of 1934 the drought continued. "When the binders, threshing machines, and combines left the fields, most Dust Bowl farmers were lucky to cut three bushels per acre. In Cimmaron County, Oklahoma, farmers harvested about the same amount as the seed wheat they planted."[89]

3

Milo Reno and the Farmers Holiday Association

Milo Reno, the leader of the Farmers Holiday Association, attracted people's attention. He was a colorful figure, possessing a deep and powerful voice, and a colorful speaking style filled with historical and biblical illustrations. His dress was also distinctive. He wore horn rimmed glasses, a bright red necktie and a broad brimmed cowboy hat. He was not a young man; he was sixty-five years old. Tall and gray haired, no longer president of the Iowa Farmers Union, but instead president of the Iowa Farmers Union Insurance companies, he was still looked up to as the elder statesman of the Farmers Union. He dramatized the farm problem as almost no one else was able to do. John Bosch, another leader in the group, was to say of him, "Reno was pretty much the firebrand that lit the fire."[1]

The fire which was to become the Farmers Holiday movement was fueled by the fear and anger farmers felt over the conditions of the Depression. In Iowa in 1932 many farmers' incomes had dwindled to half of the amount they had received a year before. Milk was selling at 1 cent a pound, hogs at 3 cents a pound, cattle for 5 cents a pound, and corn for 10 cents a bushel.[2] Prices were so bad that A.J. Johnson, a corn-hog farmer in western Iowa, received a total of 70 cents from a Sioux City packing plant for a 700-pound hog.[3]

Low prices kept farmers from making payments on their farms and equipment. In Iowa, 55 out of 1000 farms were being foreclosed in 1932. In 1933 foreclosures would rise to 78 out of 1000, the highest rate in the nation.[4] The proportion of farmers who were tenants increased in Iowa in the 1920s and 1930s. In 1920 tenants were 41 percent of all farm operators in Iowa; in 1925, 44 percent; in 1930, 47 percent; and in 1940, 49.6 percent.[5]

1933 Farm foreclosure sale in Iowa. (Courtesy of Franklin D. Roosevelt Library, Hyde Park, N.Y.)

Those who managed to hold onto their farms saw their value diminish. In 1930 the average value of an Iowa farm was $19,000; in 1935 the average value of an Iowa farm was only $11,000.[6]

Worried, unhappy and angry, farmers turned to their farm organization leaders for help and advice. Reno was well acquainted with the farmers' situation, and for nearly all of his life had worked on the farmers' behalf, working first as an organizer with the Farmers Alliance in the 1880s and supporting William Jennings Bryan in his battle for the free coinage of silver in 1896.[7]

Following his work with the Farmers Alliance, Reno operated the family farm, joining various farm groups: first the Grange and then the Farm Bureau. He left the Farm Bureau because he "was fully convinced that the Farm Bureau was organized by big business and for big business."[8] He joined the Farmers Union in 1918 and was elected president of the Iowa Farmers Union in 1921.

As a member of the Iowa Farmers Union he introduced a resolution in 1920 in the state convention demanding production costs for the farmers' products and the right of the farmer to use force if necessary to achieve them.

According to Reno the cost of production would be determined by professional economists who would also predict the amounts needed of various commodities in the national market. The federal government would then forbid the sale of the commodity for prices lower than the determined cost of production. Although his ideas on cost of production were not widely accepted by the rank and file members of the Farmers Union at this time, he continued to press for their acceptance.[9]

Throughout the 1920s he had worked for the acceptance of his ideas, and as a member of the Corn Belt Committee, he introduced a resolution in 1927 which was passed unanimously: "If we cannot obtain justice by legislation, the time will have arrived when no other course remains than organized refusal to deliver the products of the farm at less than production costs."[10]

In 1930 Reno addressed the Missouri Farmers Association at its annual meeting. He sharply criticized the Farm Board, which had been set up under President Herbert Hoover to provide cooperatives with low-interest loans and to establish corporations to hold surpluses for future sale.[11] Reno believed it intended to crush all the old farm organizations that refused to follow its commands. For example, the Farm Board had furnished funds to set up another livestock commission firm in South St. Joseph to compete with the Farmers Union firm there.[12]

Then in 1931 John Bosch, a young Minnesotan who farmed with his brothers near Atwater, Minnesota, and had been elected chairman of the county Farmers Union, spoke to that group in their annual meeting. He called for the farmers to set up an organization so that they could withhold commodities from market for the purpose of raising commodity prices. They should use this organization to pressure the federal government to pass laws to help them. If unsuccessful in their attempts then they should strike. Officers of the national organization were at the meeting and asked him to give a similar talk at the state convention. This was followed by requests to present his ideas at the National Farmers Union Convention.[13]

Reno picked up on Bosch's ideas and urged their adoption. At the national meeting, John Bosch and Milo Reno tried to get a resolution passed calling for farmers to refuse to buy or sell until farmers were receiving cost of production prices. They were unsuccessful in this attempt, however.[14] Bosch believed that the reason some of the delegates refused to go along with his proposal was because some of the state groups had substantial properties and they felt that a strike might endanger these properties.[15]

There were signs of increasing agricultural unrest. Earlier in the year, farmers in eastern Iowa had become aroused by concerns about veterinarians and the tuberculin tests being made on farmers' herds. Milo Reno, as a Farmers Union leader, heard about the "Cow War" (as it came to be called) after the first few farmers had refused to let the state veterinarians inspect their herds to determine if their cows were infected. The farmers objected to the tests because they believed that they were not valid and also because the owners were not fully compensated for the cows destroyed because of the test's readings.[16]

Furthermore, the farmers were suspicious because even though the cows were declared tubercular and ordered to be destroyed, the packer could buy the meat at a lower price. The farmers suspected that the packers and the veterinarians were cooperating to trick the farmers and cheat them. And the farmers could lose $50 to $300 per cow, and much more if a bull was ordered destroyed. This, coming on top of the economic problems of the times, turned ordinarily law abiding citizens into resisters.

Following the first few incidences in Cedar County, Iowa, a large number of farmers traveled to Des Moines, the Iowa state capital. A group of about 500 farmers arrived on a special train from Cedar County. Two thousand more farmers came from other parts of the state. They were generally Farmers Union men, but had decided to form a special organization, the Farmers Protective League, so that the Iowa Farmers Union property could not be sued by some angry veterinarian or state agent if crowd action got a little rough.

As they marched through the streets of Des Moines they carried signs conveying their sentiments. One sign read:

> Fake, Fake, Fake
> Vets condemn our cattle
> And to the packers take
> Fake, Fake, Fake
> We oppose compulsory T.B. tests
> We demand justice.[17]

The members of the Farmers Protective League came to see Milo Reno. Reno had long been suspicious of university training, especially because the county agricultural agents, usually university trained men, worked with the rival Farm Bureau. He told the farm leaders packed into his office that their cause was just. They were defending freedom, not only theirs but the freedom of others in the nation. They were showing the independent spirit of the Revolutionary Minute Men. Already angry, the farmers were stirred even further by Reno's counsel.[18]

The group left Reno's office and moved up the hill to the Iowa state capital building. In the House chamber they were permitted to address the representatives. They told the legislature: "We are here to demonstrate against compulsory tuberculosis tests and to urge the passage of the House bill to make testing optional." J.S. Stamps had a poem to read:

> Knights once went forth with lances
> Clad in coats of mail
> Now they go with squirt guns
> And shoot cows in the tail.

Throughout the meeting the farmers called for Dan Turner, the governor of Iowa. Dan Turner was a middle-aged farmer-governor and sympathetic to the plight of the farmers, but he felt that it was his duty to enforce the compulsory tuberculin testing law. He refused to give in to the farmers' wishes.[19]

When the veterinarians, accompanied by some state law officials, returned to Cedar County to test more cows, they were met by a crowd of farmers who carried them off the farmers' land and sent them on their way to Des Moines.

For a while no new herds in Cedar County were tested, but in the middle of August the agents returned to complete their testing. Farmers tried to delay their tests by not being at home when the veterinarians came; a few farmers hired their own veterinarians. Then farmers began to retaliate. Art Fogg met the agents with a shotgun, and his wife and daughter threw eggs and emptied chamber pots on the heads of the government agents. Bill Butterbrodt kicked a veterinarian in the back. Both Fogg and Butterbrodt were arrested, but not, apparently, their wives or children.[20]

Governor Turner had decided to go to Washington to plead with Hoover once more for help for the farmers during the Depression. In later years Turner would recall, "Three times I took our problems to President Hoover, three times he refused to face it. In fact, he would not acknowledge there was a farm problem. He would lean back in his chair after I told him conditions in Iowa and say, 'This is not what I hear.'"[21]

Turner appointed Joe Newell, a former police chief in Des Moines, to deal with the obstinate farmers in Cedar County. Newell recruited 80 men — state agents, sheriffs and deputies— to enforce the tubercular testing rules.

The news reached Cedar County, and approximately 500 farmers gathered at Jake Lenker's farm to resist the state officers. Tear gas was fired at the crowd, but instead of dispersing, the farmers fought back. They used clubs, bricks and bare knuckles. The assistant attorney general was thrown in the horse watering tank. Government forces beat a strategic retreat, abandoning many of their vehicles at Lenker's farm.[22]

Governor Turner called out the National Guard. He said, "Where men are organizing against government, there is only one thing to do, and that is to put down the insurrection. That is exactly what I propose to do in Cedar County."[23]

In the battle at Lenker's farm, some of the farmers had declared to reporters that if the Guard were called out, "there will be a revolution and 100,000 farmers will be in Des Moines next morning to clear out the statehouse."[24]

But when the 1,800 Guard members accompanied the veterinarians into Cedar County to finish the testing, there was no more serious violence. The troops pitched their tents at the Cedar County fairgrounds in Tipton, Iowa. They seized Jake Lenker and locked him in a horse stall. The Guard was on duty for several weeks and even traveled down into Des Moines county before opposition to the tubercular tests gradually faded away. Eventually, Lenker and Moore were tried and sentenced to prison for three years for their part in the Cow War of 1931, but were paroled after four months.[25]

The Cow War was one evidence of the growing unrest among farmers. Another example was the growth of agitation over mortgage foreclosure sales.

Farmers had been using penny sales sporadically during the 1920s. Ed Kennedy, a Farmers Union leader who became active in the Farm Holiday movement in the 1930s, remembered organizing an early penny sale in Grand Junction, Iowa. He called it "crowd control." He said he talked with the farmer whose goods were going to be sold to pay off his mortgage. "I said, 'Now if you had your little dairy herd, your farm equipment and everything at the end of the sale, would you be happy?' And he said, 'Oh God, yes.'" Kennedy then told him to keep their conversation quiet. Just about a half hour before the sale began, Kennedy asked five farmers he knew well to stand in prominent places on the inside of the sale ring, and when anything came up for bidding, to make a bid of three cents—and all of them bid three cents and never change it. He believed that the rest of the crowd would think they were the only ones not in on the deal and go along with it. And most of them did. Kennedy said he grabbed one man who wanted to bid higher, and told him, "If I were you ... I'd turn around and ... get the Hell out of here and I'd stay out."[26]

As hard times continued for farmers, the idea of penny sales spread throughout Iowa and into other states of the Middle West as well.

Women also played a part in the penny auctions. Bosch recalled one farm foreclosure sale. "We surely had a splendid bunch there that day. Nearly everyone had a rope or a fork, and more effectively, he had his wife along."[27]

In another example, Bosch remembered an auction where the farmers and their wives gathered. "There were all the farmers in the area. The farmers' wives. They had ropes hangin' over the limbs of the trees...."[28] Another person remembers a penny auction where the women "took off the pants of the auctioneer and locked him in the barn and left him."[29]

The crowds of farmers gathering to protect their friend or neighbor from the foreclosure sale of his farm and home were symptoms of the troubles of the deepening depression across the land. The problem of farm foreclosures struck Reno in several different ways. As president of the Iowa Farmers Union insurance companies, whose capital was invested in Iowa farmland, it was his duty to protect the investments of the farmers in their insurance company. But could he as a farm leader encourage his company to foreclose on farmers' lands?

Reno seems to have believed that the federal government, if properly pressed, might help the farmers out with loans. After all, the Hoover administration was lending money to businessmen through the Reconstruction Finance Corporation. Why not aid the farmers?[30]

Farmers at a livestock auction at the Central Iowa Fair in Marshalltown, Iowa. An Arthur Rothstein photograph. (Courtesy of Library of Congress, Washington, D.C.)

The depression continued to deepen across the nation. On February 19, 1932, a large crowd of Iowa farmers, perhaps a thousand or more, gathered at the Boone County fairgrounds in response to a call from John Chalmers, president of the county Farmers' Union. They formed a group to protest the low prices they were receiving and agreed to "stay at home — buy nothing — sell nothing." This was the first organizing action of what would become a regional, even national, farm withholding movement.[31]

By March 1932, large meetings were called for Grinnell, Webster City, Fort Dodge, Harcourt and other Iowa towns.[32] Then Milo Reno and John Bosch called for a statewide meeting of farmers to be held at the state fairgrounds in Des Moines on May 3, 1932.

In trying to think of a name for the organization, someone said, "Well, we just had a bank holiday.... Why don't we call this a farm holiday?"[33] As the movement grew, the name stuck. About two thousand farmers attended the first meeting at the fairgrounds. John Simpson attended the meeting and attacked the Republican administration. A short platform was written which called for direct action. The group pledged to organize farmers throughout the nation to withhold their commodities until cost of production prices were reached. They elected Reno as president.[34]

Some unknown farmer-poet's praise of the Farm Holiday was published in the *Iowa Union Farmer*:

> Come fellow farmers, one and all —
> We've fed the world throughout the years
> And haven't made our salt.
> We've paid our taxes right and left
> Without the least objection
> We've paid them to the government
> That gives us no protection.
> Let's call a "Farmers Holiday"
> A Holiday let's hold
> We'll eat our wheat and ham and eggs
> And let them eat their gold.[35]

There were many articles in the *Iowa Union Farmer* concerning the plans for the coming Farm Holiday strike. Reno and other Farmers Union leaders traveled throughout Iowa and the surrounding states making speeches to farm groups urging farmers to join up. They got tens of thousands of farmers to sign up.[36]

John Bosch remembers that when he called a meeting in a given area, such as Montevideo or Marshfield, "When I sent out the call, we could bring together twenty or thirty thousand people without much notice."[37]

The group that met in Des Moines on May 3 had selected a date for a farm withholding strike. It was to begin on July 4, possibly to give a patriotic effect to the movement. Milo Reno and the other officers of the association did not want to call a strike at that time. They felt that they were not sufficiently organized. Also, since prices on farm goods had risen a little, some observers hoped it signaled the end of the Depression. So it was postponed.[38]

But not for long. In every county in Iowa, and in many surrounding areas, recruiters had been talking with their friends and neighbors, urging them to join. Farmers who joined the Farm Holiday Association had been signing a striking pledge: "I hereby agree that I will not haul any more farm products to markets while the farm holiday is on. I also agree to call on all truck drivers in my vicinity and urge them to comply with this agreement."[39]

The strike began on August 8, 1932. According to Bosch, the strike in Iowa was a spontaneous kind of movement. He remembers: "I was on my way down to Iowa to attend a meeting where we were just beginning to formulate plans, when I ran into this picket line myself. That was the first I knew about it."[40]

Bosch and Reno would have preferred another time. Bosch said it began "before we were anywhere near ready to call it.... But the bitterness of the Farmers Union in that area was so great that there just wasn't any way of stopping it."[41]

In some ways it was a good time for a strike. By August the farmer's corn had reached a point where it no longer needed cultivation: The hay had been cut and put up in hay mows, and wheat threshing was nearly finished. Farmers had a brief period of time before other harvesting tasks began, so they could participate in the Farm Holiday activities.[42]

During the first few days of the strike very little happened. Then on August 10, 500 dairymen gathered at a hotel in Sioux City to present their demands for an increase in the price they were receiving from the Sioux City distributors. They were reacting to the low price they received for their milk. They were receiving two cents a quart for milk that was being sold to consumers at eight cents. Leaders of the strike movement said they could no longer afford to deliver milk at such a low price.[43]

They presented the distributors with their demands, but the distributors refused to raise the prices they paid to the farmers. After receiving the distributors' answer, the farmers formed an association, the Sioux Milk Producers' Association, and ordered a milk strike to begin the following day.

The roads into Sioux City were almost immediately picketed by the dairy farmers. All trucks going into the city were stopped, and nearly all of the trucks carrying milk were turned back. Dairy farmers in South Dakota also participated in the Milk Strike. The picketers leaped on the running boards of farm trucks or threw hay bales or logs in the path of oncoming vehicles.[44]

At first the mood of the pickets was hopeful. The farmers believed that the strike might accomplish something. They only expected the strike to last a short time. And because the Association had established free milk depots for consumers, and milk was being delivered to hospitals and to orphanages, they felt that no one was really being hurt.[45]

Women were on the picket lines too. Papers reported that "hundreds of farm men and women placed barricades in roads to prevent produce-laden trucks from entering Sioux City." Other accounts mention 1,500 "farmers and their wives" patrolling roads and blocking highways with their cars, or dragging logs across roads to stop truckers from going through.[46] Farm women also brought food and coffee to the picket lines.[47]

In these groups of pickets there were also numbers of young men, some of them teenagers from the towns as well as from the farms, perhaps drawn to the picketing by the novelty and excitement of the occasion. They seemed to believe that they were participating in a struggle to bring justice to the farmers. Some observers feared that these young men were more likely to use violence than the older, experienced farmers.

From Kingsley, Iowa, came reports that 400 farmers had visited all the produce dealers in the town and warned them not to open for business or they would have all their produce thrown into the streets.[48]

The farmers soon learned that not everyone was in favor of the Farm Holiday action. Almost immediately, truck drivers tried to crash through the blockades set up by the striking farmers. Only four days after the strike began, drivers were met with violence as a crowd of about 400 farmers dumped a load of milk on the highway.[49]

Blockade runners were sometimes non-striking farmers, or employees of distributors. Governor Turner recalls that in the early days of the strike he talked with a farmer in southwestern Iowa who fed large numbers of hogs and also cattle. They talked about the Farm Holiday strike, and Turner asked him, "What are you going to do about it?" The farmer answered, "I have six loads, about sixty head, of fat steers ready for market. I have arranged for trucks to send them [to market] tomorrow. I will have on each truck two of my neighbors with shot guns, and we propose to take them to the stock yards and no one better try to stop us."[50]

The pickets set up more formidable defenses. The most powerful defense was to lay large 4"×4" planks, 16 feet long and studded with spikes, across the highways at strategic spots. This was very effective. Turner recalls that he heard of no truck passing over these spiked planks.[51] Farm pickets grew in size, sometimes up to a hundred or more.

The pickets had an effect; they were cutting off the flow of milk and produce into Sioux City. The strike was also spreading to other cities, to Council Bluffs and Omaha.

Police retaliated. The sheriff of the county surrounding Sioux City advertised for a hundred deputies.[52] The sheriff at Bluffs promised, "I'm going to keep the highways open if I have to deputize a regiment."[53] At Council Bluffs, hundreds of strikers, "including women and children," were tear gassed by police officers who drove through a crowd with cans of compressed tear gas fastened on the running boards of their car. Three Farm Holiday women were injured on this occasion.[54] When the police officers began their return trip through the crowd, the strikers gathered about the car, throwing bricks and clubs, one of which crashed through the closed windows of the car, hitting one officer and resulting in cuts and bruises to the others.[55]

On August 25 police deputies seized Farm Holiday picketers along the roads and took them to jail. Forty-three men were put in jail in Council Bluffs.[56] The news spread through western Iowa. Milo Reno was questioned by reporters about the trouble in Council Bluffs. He answered that he had not heard about it. But he called Governor Turner and told him he ought to get the farmers out of jail, and also suggested that he call the governors of the Middle West together for a conference.[57]

Farmers from all over western Iowa got into their cars and trucks and headed for Council Bluffs. Governor Turner received word from the sheriff of Pottawatamie County that several hundred farmers were traveling to Council Bluffs, from at least as far north as Sioux Falls. They were being joined by other farmers as they came south. The sheriff had heard that the men were determined to break into the jail in Council Bluffs and liberate the men who had been jailed. Turner said, "I can send you as many soldiers [National Guard] as you think you will need to prevent the strikers from breaking out the prisoners." The sheriff said he would call back in two hours. When he called back he said he thought he could deal with the farmers if he could deputize fifty or sixty men to hold back the strikers. Turner remembers that he cautioned the sheriff not to arm the men.[58]

There were detectives brought in from Omaha, and machine guns mounted. Estimates are that there were nearly 3000 Farm Holiday men who surrounded the county jail and faced the sheriff's deputies. The farmers were angry about the women and children who were fired upon. They muttered remarks about the rights of free men, and the need for a revolution. Some men in the crowd of farmers threatened to storm the jail.

But at this point a farmer asked to be allowed to pass through the police lines. He thought he had a solution to the problem. He offered to put up his property as bail for the prisoners. The sheriff accepted his offer and the prisoners were released. The crowd of Farm Holiday men dispersed peacefully, averting a situation which could have ended in violence and bloodshed.[59]

Another incident had more serious consequences. Shots were fired at a group of picketers near Cherokee, Iowa. Fourteen of the farmers were wounded. The county sheriff and two other men were later put on trial, but none were convicted.[60]

Governor Turner was receiving hundreds of calls and letters from Iowans demanding that the roads be kept open. He hesitated to call out the National Guard because there was a possibility of more bloodshed in fighting between the deputies and the young men on the picket lines.[61]

Eleven police officers were hurt in a clash with Farm Holiday picketers near Cushing, Iowa, August 29, 1932. There had been a fight between the picketers and 50 deputies who were attempting to escort five truckloads of hogs for a farmer who lived near Moville to the county lines. The deputies said they were attacked with pipes and sticks.[62]

In James, Iowa, deputies from LeMars, armed with shotguns, rifles, revolvers, billiard cues and baseball bats, escorting twenty-two truckloads of livestock, were halted by a crowd of 400 to 1,000 strikers. The sheriff ordered his men to remove the barriers and announced that he would shoot anyone who interfered. Nevertheless, the farmers rushed the deputies and forced them to climb back into their trucks. As the trucks turned around to go back to LeMars, the strikers threw stones, bricks and other missiles at them.[63]

Although Reno talked strongly about keeping the strike going, the fights between farmers and truckers worried him. Concerned about the escalating violence of the Farm Holiday strike, Reno declared the strike a "truce." He said, "We will not jeopardize the lives of unarmed farmers."[64]

He also hoped that others might find a way to help them achieve their goals. Governor Warren Green of South Dakota was asked by the Davison County Farm Holiday Association to call a governors conference. It was hoped that the governors might establish a means for preventing farm produce from reaching the markets "until a permanent scale of prices shall be dictatorially fixed or established of a sufficient amount to give the farmers a fair exchange value for the products of their industry on a parity with that of any other American industry."[65]

Governor Green conferred with several other governors, and they agreed to meet at Sioux City on September 9 and 10. Green said that the meeting should be strictly non-political, considering solutions rather than causes of the Depression.[66]

Emil Loriks, a leader of the South Dakota Farmers Holiday Association, said that they would hold up the proposed strike in South Dakota until after the Governor's Conference. Meanwhile, he and E.H. Everson, president of the South Dakota Farmers Union, continued to actively recruit and organize the Farm Holiday Association in South Dakota.[67]

Not all the farm leaders wanted to postpone action or agreed that the conference should be non-political, however. Ella Reeves Bloor, a worker for the Farmers Union League of North Dakota, hurried to Sioux City to be part of the activities there. She often spoke to large groups of farmers and reporters. One of the groups has been estimated to have been as large as 500.[68] She declared, "It lies in the power of the workers and farmers of America to change this country and do as the workers of Soviet Russia have done." In a question period following her speech, Mrs. Bloor declared that the governors of the mid-western states who were coming to Sioux City "have no power to do anything and don't want to do anything. They're elected to serve a certain class. I think they'll try to stop the strike, but I believe the farmers will stick together."[69]

In an interview with reporters, Reno suggested that the governors declare an embargo on the movement of farm produce at prices below cost of production beyond the borders of their states.[70]

Present at the conference were Governors Turner of Iowa, Green of South Dakota, Olson of Minnesota, and Shafer of North Dakota. Also attending were representatives of the governors of Wisconsin, Oklahoma, Ohio, Wyoming and Nebraska. To greet them, approximately 5,000 Farmers Holiday members marched through the streets of Sioux City.[71]

The governors met for two days and listened to the representatives from the various farm groups as they described their problems. All the groups who came protested the low prices of farm products, and complained about failing banks and unemployment. Reno presented a series of proposals: state mortgage moratoriums, a special session of Congress to enact the Frazier farm credit bill, voluntary action by farmers to withhold their produce from markets, and state enforced embargoes against the sale of farm products at prices less than the cost of production. His statement of demands was signed by Farm Holiday leaders.[72]

Turner replied to Reno's request for a state enforced embargo on farm products by saying he was asking for the impossible. "If you insist on something that strikes at the very root of our government, you will destroy only our chances of getting results."[73]

John Bosch said that the low price level of farm products was an evil striking at the foundations of government. "It will be impossible to avert a revolution — our homes must be saved or there will be a revolution," he told the governors.[74]

The conference leaders put in a long day, hearing testimony from 40 witnesses. As the meeting wound down, representatives of the Farmers Holiday stated that the Farm Holiday strike would continue unless the governors' action was satisfactory to them. Turner replied that he would have a statement ready for them.

The next day the governors and representatives signed Turner's document. It stated: "We have been urged to place an embargo on the shipment of farm produce out of the states here represented ... [but] we are obliged to deny the request." The request was denied because there was no legal precedent to such action, and because it would cause many "complications."[75]

Governor Turner remembers that many of the strikers were bitter at the stand which the governors had taken. But they took no action and left that evening for their homes. He wrote, "I realized that nothing concrete had been accomplished. There must be some other solution. If I did not find one, I would be compelled to call out the National Guard. This I did not want to do, but I knew the roads must be opened."[76]

The businessmen of Sioux City protested the blockade movement, which they said was costing businesses in their city thousands of dollars.

When questioned at his home by reporters, the governor leaned out his window and shouted to the reporters that he had a plan in mind to settle the affair without troops, but he declined to disclose it.[77]

Turner decided to call two men who were active leaders in the Holiday movement. They were "good, sturdy citizens who became desperate because they had lost all they had and thought there was only one recourse and that would be violence." They agreed to meet with him. Turner told them that he was determined to open the highways. "Even if I had to do so the hard way, namely, by using the Iowa National Guard." Turner said that all his life he had fought for justice for the farmers. He knew that he had been elected because of the backing of farmers and small town businessmen. But he could not allow mob rule to prevail. And, he reminded his listeners, "You fellows aren't the whole show, a lot of farmers want to get to market."

It worked. "On Tuesday the roads were opened. Two influential and essentially good, patriotic citizens got the job done, as I had hoped and thought they might do."[78]

An announcement from Milo Reno was published in the newspapers the next day, saying that he had decided to call off the Farmers Holiday strike. Turner believes that Reno was glad when the strike collapsed. "It was too strong a medicine for him."[79]

Some farm leaders were happy to see it called off. In Missouri, William Hirth, president of the Missouri Farmers Association, wrote that he did not believe that the "farmers strike" could achieve its objective by peaceable means. He wished he could support it because "the farmer's position has become utterly desperate."[80]

The reasons Hirth believed the strike would not be effective were because Congress needed to deal with national issues, such as the tariff to raise prices; states could not deal with prices individually, and farmers could not continue to withhold from market perishable goods such as milk, butter and eggs without themselves suffering in the process.[81]

What were Governor Turner's feelings about the Farm Holiday? Why did he not call out the National Guard immediately to stop the picketing of Iowa roads? He said it was because he feared bloodshed. Others suggested that he might also have had another motivation; he wanted to get re-elected.[82]

In South Dakota the Farm Holiday declared a strike. Emil Loriks explained the movement's program in a half hour radio talk. He said that they planned to withhold non-perishable farm products from the market for 30 days, and if at the end of that time prices did not equal the cost of production, farmers would be asked to withhold perishable products also. Picketing was not to be part of their program. Loriks said, "It is the purpose of the association to carry on a peaceful strike and to avoid a situation like that present at Sioux City."[83] However, little tangible results were seen from the first few days of the strike.[84]

In Minnesota small Farm Holiday strikes were organized near Pipestone, Montevideo and Willmar.[85] As the strike spread to other parts of the state, including St. Paul and Minneapolis, so also did the opposition to the strikers. Crowds gathered at the picket points, and Farm Holiday members and truck drivers engaged in hand to hand fighting. In Anoka Mrs. Margaret Heldt used a hammer on the toes of picketers as they attempted to stop her truck. The Sheriff of Hennepin County said they might have to appeal to the militia: "Things look pretty bad."[86]

In North Dakota A.W. Ricker told the annual Farmers Union Convention that the only means by which the farmers would be able to obtain their prices for their commodities would be through affiliation and cooperation with the Farmers Holiday movement.[87]

Not having been particularly successful in their withholding action, the Farm Holiday Association turned their attention to stopping farm foreclosures. They used local councils of defense, made up of five to nine members. Milo Reno's "Instructions for Organizing Farmers' Holiday..." were passed around widely in the Middle West and used by various councils in determining their strategies.[88]

John Bosch recalls that, as far as Minnesota was concerned, he could stop almost any sale with a telephone call if he felt that the sale would be unjust to the farmer. But he occasionally had to resort to subterfuge to convince a reluctant sheriff to halt a sale. For example, he would station some people outside of the building and then go inside to talk with the sheriff. If the sheriff refused to give in, then someone outside of the building, or around the corner, where he could not be identified, would call out, "Let the so and so out. Cut a hole in the ice. We'll put him down twice and pull him up once." Or somebody else would say, "Like hell! I've got a rope waitin'," and so forth. "And he thought it was real," said Bosch. "And it scared him." And then Bosch would suggest that the sheriff call his attorney again, and after a call or two the sale would be called off.[89]

If these methods did not succeed, the Farm Holiday leaders might try a "penny sale." In Nebraska they were called Sears Roebuck sales. Harry Lux, who was instrumental in organizing many of these sales, recalls that an auctioneer told him that they were legal and the courts had recognized them as valid.[90]

In South Dakota the threat of penny sales was so frightening to creditors that few sales actually occurred. However, Holiday leaders worked to encourage banks and other lenders to renegotiate farm loans so that farmers would have a longer time to pay them off. Emil Loriks of Arlington, South Dakota, as a leader in the Farm Holiday movement, traveled thousands of miles meeting with defense councils and working to find acceptable settlements for all the parties in mortgage disputes.[91]

Loriks generally tried to avoid the use of penny sales if possible. But he heard about the sales that did happen. He was told about one notable penny sale which occurred near Milbank, South Dakota. Approximately five hundred farmers gathered for the sale, and also the county sheriff and some of his deputies. When one of the deputies fired his gun, the farmers surrounded the sheriff and his deputies, took their guns away and told them to leave. After that the farmers held a penny sale and then returned the goods to the farmer.[92]

Governor Turner called them "plugged sales." He writes that he knew about them. "Of course, as Governor I had all the information and knew these auctions were being rigged. I did not attempt to stop them. No law was being breached...."[93]

Turner, a Republican, went down to defeat along with President

Hoover and many other Republicans in the November 1932 elections. Farm people looked to the new administrations, under Democratic president Franklin Roosevelt and Democratic governor Clyde Herring, for relief from the problems of low farm prices, mortgage foreclosures, etc.

In November the Farm Holiday national board decided not to hold a general strike until the incoming administration had a chance to inaugurate a new program. But they warned that if it were similar to the "thumb twiddling" program of the Hoover administration, they would call a national holding movement.

In a letter to a farmer in Missouri, Reno wrote: "We are waiting patiently for Governor Roosevelt to have an opportunity to carry out his pledge to the farmer...."[94]

John Bosch believed that the Farm Holiday resistance to foreclosure sales led to state and federal action. He said that Minnesota Governor Floyd Olson called him in and "said, 'Now look, John ... as far as the state is concerned, the state district judge and the country sheriff, I can pretty well protect you. But you can't buck the U.S. Government.' Well, I said, 'Floyd, we did.'" Later, many states declared a moratorium, and the federal government passed the Department Conciliation Commissioner Act. The governors of the states would recommend a conciliation commissioner to the President and he could appoint them.[95]

Milo Reno continued to write editorials in the *Farm Holiday* newspaper. He wrote about his opposition to the New Deal's proposed farm program: "I am unalterably opposed to the Allotment Plan for many reasons, but, especially do I oppose any proposed legislation for the American farmer that does not recognize his right to production costs, which the Allotment plan does not do; in fact, it has carefully avoided even mentioning it. Any agricultural legislation that does not recognize the farmer's right to production costs is evading the issue in the interests of the farmer's exploiter and can be of no permanent value to the farmers of the nation."[96]

When Reno learned that President Roosevelt had nominated Henry A. Wallace as Secretary of Agriculture, he wrote in his editorial that, although he would have preferred John A. Simpson as Secretary, he would be fair and support Wallace in "every laudable effort to correct the [agricultural] situation." But he warned, "This pledge of support does not mean a slavish acquiescence in whatever program that Secretary Wallace may propose or support."[97]

Emil Loriks wrote Reno in April 1933 that he was in Washington contacting various leaders concerning the proposed legislation. "Personally," he wrote, "I am confident that whatever progress we make under this new Administration will be due primarily to the efforts of the F. Union and the

Farmers Holiday Association. Believe they are more scared of the Holiday Association than anything else right now."[98]

As the House and Senate were discussing the New Deal farm plan, Reno received several letters from congressmen assuring him that they were in sympathy with the attempts by the Farm Holiday and Farmers Union to amend the administration's program. G.M. Gillette of Iowa wrote that he was "heartily in sympathy with the amendment passed by the Senate, providing for production cost of farm products." Although there were attempts in the House to prevent amendments to the bill, he would "continue to oppose all efforts to gag the membership." But he feared that they did not have enough votes to prevent this.[99]

Another congressman wrote Reno that, in addition to refusing to permit amendment to the farm bill, the administration refused to permit amendments to the farm loan bill, which would have reduced the rates of interest and increased the amount of the loans. "The leadership in control were able to hold their membership and thereby defeated an opportunity to liberalize this bill."[100]

Then came an incident which illustrated the possibilities of violence inherent in the attempts to prevent foreclosures. In April 1933 a group of farmers near LeMars, Iowa, attempted to stop the eviction of a farmer from a farm he had been renting. They went to see the administrators of the estate that owned the farm and were unable to achieve a compromise. Then they headed for the courthouse at LeMars. At the courthouse, Judge Bradley reprimanded the farmers for entering his courtroom wearing their hats and smoking. The already irate farmers seized the judge, saying, "This is our courtroom, not yours." They dragged him from the room and ordered him to promise not to foreclose or sign any more foreclosure notices. He refused and was slapped. After again refusing to go along with their requests, he was again slapped. Then he was taken to a crossroads out of town and a rope was put around his neck. The farmers started to hang him from a telephone pole. But then there was some debate among the group as to whether it was better to hang him or drag him behind a car. They eventually removed his pants and filled them with grease and dirt and then left him. He returned to his home with rope burns around his neck and bloody and battered lips.[101]

The National Guard was called out the next day. Andrew Ball, the attorney for Crawford County, said he thought most of the people involved in the LeMars affair were members of the Farm Holiday: "The farmers were holiday sympathizers, organized to stop foreclosure sales all over this part of the country."[102] Claiming that they knew who led the assault on Judge Bradley, the guardsmen began making arrests soon after their arrival.

Milo Reno's remarks on the LeMars incident showed his concern and mixed feelings: "It is deplorable, in fact revolutionary, when people who are law abiding, conservative citizens ignore the courts and violate the law even to the extent of mobbing judges; however, it has occurred a great many times in the world's history...."[103] The men who were arrested for their part in the LeMars incident were eventually tried and given light fines and sentences.[104]

Also in April 1933, the Farm Holiday leadership began to plan for an anniversary meeting of the Farm Holiday. It had been a year since the group was formed. Because the Iowa Convention had voted in favor of a farm strike by May 3, 1933, if Congress had not assured the farmers of cost of production, there were rumors that this national meeting of the Farm Holiday would also initiate a new strike.

Reno busied himself making arrangements for the May 3 meeting. Farm Holiday members from New York, Maryland, Ohio, Texas and New Mexico wrote that they planned to attend the meeting.[105] News was also received that a delegation from the Canadian province of Saskatchewan were planning on coming.[106]

Mrs. Algot Johnson of Frederic, Wisconsin, wrote, saying that she and other members of the Farm Holiday movement from Wisconsin would be coming by truck to Des Moines. She asked if it would be possible for the women to stay in a farm member's home, since the men would likely stay in tents. She hoped Reno would reply, although she knew that "you are one the most busy men in our country."[107] Reno replied, "You Wisconsin people surely have the right viewpoint and the right kind of grit. While I live in the city and have quite a family, I assure you that we have room for you and your friend.... I only wish I had accommodations for all the ladies from your county."[108]

Another farmer wrote: "A number of us are planning on coming out there May 3rd for the Holiday meeting. We can take our own blankets along, but as mattresses are rather bulky I'm wondering how chances would be to 'borrow' some farmer's straw stack or hay mow for a night or two so our expenses would not run too high."[109]

Reno wrote a Nebraska farmer: "The meeting of May 3rd to be held in Des Moines ... will determine our program for the coming year, and I have every reason to believe that this group of men ... will unanimously decide to either have justice or strike."[110]

On May 3 the National Farm Holiday meeting was held. There were delegates there from seventeen states. They listened to speeches, sang songs and voted unanimously to call a strike on May 13, 1933. The assembly adopted the following resolution: "That the committee recommends unanimously that the National Farmers' Holiday Association proceed to declare its marketing holiday on all farm products, May 13, 1933, and that its original legislative demands be again presented to congress...."[111]

In the May 9 issue of the *Iowa Farmers Union* paper, Reno wrote that he had received word that the cost of production feature had been removed from the agriculture bill as it was being considered in the joint committees of Congress. Reno commented in despair, "This is certainly the last straw necessary to convince the farmer that the program of the future is to reduce him to absolute serfdom...."

Reno sent a letter to Secretary Henry Wallace stating that the farmers were not asking for any favors or considerations that had not already been granted to other groups in society. What the farmers wanted, what they believed would solve their problems, was receiving costs of production for their farm produce.[112]

That Reno was thinking seriously about the problems and complications of calling a nationwide strike is evidenced in a long letter he wrote to Cashman: "I am very much concerned lest we blunder seriously in our Holiday movement, but even though we do blunder, we have at least tried."[113]

Reno sent a telegram to President Roosevelt on May 11: "ACCORDING TO PRESS REPORTS YOU ARE WILLING TO DO ALL IN YOUR POWER TO AVERT FARMERS' STRIKE AND RESULTANT CONFUSION STOP WILL YOU DECLARE MORATORIUM ON FARM FORECLOSURES AND EXECUTIONS UNTIL FAIR PRODUCTION COSTS ARE CONCEDED FARMER ANSWER IMPORTANT."[114]

On May 12 President Roosevelt signed the Agriculture Adjustment Act. It was described as a bill to raise farm prices, giving him inflationary powers and providing for the easement of the agricultural mortgage burden. However, the bill did not provide for cost of production legislation, which the Farm Holiday and Farmers Union had been seeking. Instead, it provided for production control, farmers limiting what they produced.

Reno was meeting with Farm Holiday leaders in St. Paul on that day. John Bosch remembers that Reno had grave doubts about calling the strike. Reno had called on Iowa's Governor Herring, who had told him that if the Farm Holiday tried picketing the highways in Iowa he would do whatever was necessary to stop it. And Reno did not know what to do in connection with this. He was looking for a way out.[115]

Bosch did not favor the idea of going on strike. "I didn't think we had any opportunity of being successful in a strike." So Bosch remembers that he talked it over with Governor Olson, and "the stage was set so that we were to contact Floyd Olson and he was to give us the reason for calling it off."[116]

Then, on the evening of May 12, Reno sent telegrams to many of the state and local leaders: "In view of Roosevelt's farm statement, at Governor Olson's request and leaders' advice, the executive committee declares a truce and suspension of the National Holiday called for Saturday, May 13th to a later date, date to be determined on the effectiveness of federal legislation...."[117]

In a letter to the presidents, Reno stated his reasons for the postponement. He said President Roosevelt had made promises to the American farmer in his 1932 campaign. Since then, the President, in his conference with Minnesota's Governor Olson as well as in his statements made at the signing of the Agricultural Adjustment Act, had "reaffirmed these pledges to the American farmer. He is entitled to sufficient time to demonstrate, first if his program is workable and second, if the law as passed ... will be carried out in ... [that] spirit...."[118]

Although Reno hoped to continue his efforts to increase the strength of the Farm Holiday movement, he was hindered by lack of funds. Another hindrance to more active organizing efforts was an attack by a "bolter group" of the Farmers Union of Iowa upon the Farmers Union Life Insurance Company, which Reno, as president, had to answer. Reno wrote that he was sorry to miss his scheduled meetings in Wisconsin, but "it required all my attention to get this matter straightened out...."[119]

Meanwhile, he was observing and trying to decide what the effect of Roosevelt's New Deal program would be. In regards to Roosevelt's agricultural program, Reno had mixed feelings. He wrote, "I am not satisfied with either his agricultural measure that ignored the right of the farmer to production costs, neither am I satisfied with the program as to refinancing homes and farms, yet I realize that their usefulness will be largely determined upon the sincerity of the men whose duty it will be to administer those measures."[120]

He was delighted, though, with the move of the Roosevelt administration to take the country off the gold standard. In an editorial in the *Iowa Union Farmer* Reno wrote: "No one doubts now that BRYAN WAS RIGHT." Reno went on to say that, although Bryan did not live to see his dream come true, he would have been pleased to see a Democratic president ask congress to "carry out the things he so long and earnestly advocated."[121]

By June, Reno was beginning to receive word from various farm groups that they were unhappy with the New Deal Farm Program.

One woman farmer who had been farming for forty-four years said that during the last nine years she had carried a heavy federal loan on her farm and gone through the "bitter experience" of threatened foreclosure and loss. She believed that she knew from personal experience what she was talking about: "I know whereof I speak."[122]

F.M. Breed of Onawa, Iowa, wrote that he heard "all up and down the line" that Reno should not have called off the strike.[123] Mrs. Charles Wedholm, who farmed on the outskirts of Duluth, Minnesota, wrote, "Please call a national milk strike before we starve to death. We have been waiting for Sec. Wallace. But we believe he is paid off and going to sleep on the job."[124]

Reno continued to organize and to make arrangements either for himself or for others to speak at farm meetings. He wrote John Simpson in June that there were so many requests for him that he hoped Simpson would send him a list of his dates and how many meetings he could "donate to the Middle West."[125]

He also began to receive discouraging news from some states. Charles Ray of Fairland, Indiana, wrote that, although the farmers had been excited about the Farm Holiday in the spring, "then the Roosevelt dream brought hopes, lack of interest, meeting[s] that I held after that was all in vain, there was a feeling that we did not need an organization to help us, that the administration was going to cause a rain of money to fall upon the farmer's head, and nothing I could do would cause them to see otherwise."[126]

John Bosch wrote, "I have been very much disappointed in the financial support that we have received in the past month or two. Nearly all of the state directors have been compelled to pay expenses from personal income."[127]

In reply, Reno wrote, "I think I understand and sympathize with the situation in Minnesota, as well as every other state. It seems to be entirely impossible to get any financial support for state and national organization." Reno went on to say that the *Farm Holiday* paper was due to be published again in ten days, and he did not know where the $200 necessary to put out one issue was to come from.[128]

In a later letter to Bosch, Reno wrote that he was really discouraged in the responses to requests to the state organizations for payment of the national dues. He did not see how the national organization could continue to function unless they received more support.[129]

Commenting on the situation, Reno wrote an editorial for the *Iowa Union Farmer*, "The Future of the National Farmers' Holiday Association." In the article he stated that the most dangerous enemy of the organization is the "disposition of men, especially farmers, to relax their vigilance ... and to neglect ... the organization they built."[130]

Reno was also discouraged with a corn and hog conference held in Des Moines in July of 1933. As he observed the conference he became angry. He believed that the same old group that had supported the Hoover agricultural program was returned to power in the new administration. On the other hand, the Farmers Union and the Holiday Association, as well as the Farm Clubs of Missouri, were treated as "undesirables."

F.W. Murphy of St. Paul, Minnesota, wrote that nothing had been done to increase the price of farm commodities. But the prices of industrial products had been rising steadily. "The farmer is still the goat, only now more so than ever." He believed that if wage scales could be established, if the hours of labor could be set, if the prices of industrial goods could be stabilized at high levels, then the government ought to be able to establish the prices of farm commodities high enough so that farmers could make a living.[131]

Reno, Condon, Kennedy and others worked for some time preparing a national agricultural code. Reno wrote a friend in Estherville, Iowa, that he wished the Farm Bureau and the Grange would support the Farmers Union and the Holiday in their demand that they be treated the same as other groups. "However, we expect to bring it to a show-down and demand that the farmer shall have a right to determine his prices, the same as [other groups]...."[132]

After the code was developed they hoped to present it to President Roosevelt, bypassing Secretary Wallace. "We do not hope for any assistance or encouragement from Secretary Wallace." If the president did not accept the code, Reno commented, "God help us!"

On September 1 Reno sent a telegram to President Roosevelt: "Bankrupt prices of farm products and failure of farm refinancing program have exhausted patience of farmers. Must have permanent corrective measures for agriculture immediately. Secretary Wallace's opposition to inflation, 'because it tends to advance prices of farm commodities' has entirely destroyed farmers' confidence and hope for any relief under his administration. Sentiment for nationwide strike growing daily. Urge you act immediately for farm price stabilization based on production costs."[133]

Reno wrote Richard Bosch that he was receiving letters in the mail from nearly every state in the Union saying they were ready to strike. "This is encouraging in one way, that our farmers are not ready to lie down and take their licking...." The Farm Bureau of Iowa, as well as the Farm Bureau of nine counties in Ohio, had adopted resolutions favoring production costs. And the governor of Iowa had wired President Roosevelt that the farmers were impatient and dissatisfied with the agricultural program, and that the National Recovery Administration would fail unless applied to agriculture, and the price of farm products pegged at production costs.[134]

In an editorial in the *Iowa Union Farmer*, Reno explained that, following the state meeting of the Farmers Union, there would be a joint meeting of the National Board of the Holiday Association and the official family of the Farmers Union. The meeting had been called to protest the administration's policy of applying the principles of the NRA to labor and industry while "ignoring the farmers' right to determine for themselves a code of ethics as well as prices.... Without production costs the farmer cannot survive, except as a beast of burden to serve the industrial and capitalistic groups, and future history will not condemn him for fighting for the liberties that he has helped to maintain for others."[135]

In his editorial in the *Farm Holiday News*, September 14, 1933, Reno discussed the telegram which he had sent on September 1 to President Roosevelt. He said that the telegram had been ignored by the president, and his secretary had referred it to Secretary Wallace for an answer. Wallace had not answered it either. And while Reno had not "lost faith in President Roosevelt's good intentions," he felt the president had not shown the "wisdom and consideration" that farmers had expected, especially in the appointment of Henry A. Wallace as Secretary of Agriculture.[136]

Now, five months after the strike had been called off, Reno said he could not see that any part of Roosevelt's program would mean a permanent solution for the farmers' problems. "Frankly, I am yet undecided as to whether this is a part of the Roosevelt program, or whether his inefficient economic idiots known as the 'brain trust' have, in a way, deceived him as to actual conditions."

Reno prepared petitions asking for the removal of Wallace from his position as Secretary of Agriculture. He asked that various farm leaders also get up some petitions, circulate them and then mail them to the Farm Holiday office in Des Moines. Reno was planning to have a committee appointed that would personally deliver the petitions to President Roosevelt.

From M.L. Glarum in Montana came another letter detailing the farmers' problems: "I thought maybe you would like to know how us farmers are getting along out here in Mont. Things are worse here than they were last fall. The grasshoppers are eating up all of our grass so we have to sell our stock, and we only got 5 bushels of wheat to the acre, and only getting 58 cents a bushel for it, and flour is selling for $2.35 for 49 lbs."[137]

Glarum concluded his letter by saying that they were planning to have a farmers meeting in another week. Although only a few farmers in the area belonged to the Farm Holiday Association, "I think if you could let me know how we can get any aid by being a member, that I could induce many farmers to join."[138]

Jess Sickler, Secretary of the National Holiday Association, passed on a letter from Mrs. Chris Linnertz, the secretary of the North Dakota Holiday Association. She wrote that farmers in North Dakota had been hit hard by the grasshoppers, and they were receiving so little income that "we can hardly buy postage stamps." She believed they wanted to strike because they were getting "impatient with conditions as they are, and they want action."[139]

A wire from Fred Schmidt of Sioux City, Iowa, was most emphatic: "Farmers demanding action if Wallace can't help them by the eternal God they will help themselves."[140]

About this time Reno received a letter from the Khaki Shirts of America suggesting that Reno and the Farm Holiday join them in a march on Washington. Evidently fearing that the group was fascist, Reno replied that he had never been in favor of a dictatorship, and in fact would do everything in his power to prevent it. Until he knew more about the group and their purposes, he refused to join in their march.[141]

As the time moved closer to the state meeting of the Iowa Farmers Union and the national board meeting of the Farm Holiday Association, Reno thanked Condon and Kennedy for the work they had done on preparing a code for agriculture. A committee of the Association was also going over the suggested code, and it would be presented at the national meeting.[142] Then he expected they would issue an ultimatum to President Roosevelt that the farmers be given the same consideration as other groups under the NRA.[143]

At the meeting of the National Farmers Holiday Association there were about 1,200 representatives from 18 different states attending. The agriculture code, which various members of the Association had been working on for several months, was presented. It called for a guarantee of the cost of production for agricultural products, a maximum work day of ten hours, setting of minimum wages by the American Federation of Labor, collective bargaining, prohibitions against child labor, and the licensing of buyers of farm commodities. The group adopted the proposed code and chose three men to present the code to the President. If the President failed to comply, the group voted to "withhold all farm products from the market." Petitions demanding the resignation of Secretary Wallace were circulated at the meeting.[144]

Reno sent a telegram to President Roosevelt: "Committee of three authorized by National Farmers Holiday Association yesterday to present farm code to President of United States. Will arrive in Washington

September twenty-eight. Requesting audience with you. National Farm Holiday strike held in abeyance pending acceptance of this code."[145]

The committee of three appointed to bring the farm code to Washington were E.E. Kennedy of Kankakee, Illinois, national secretary of the Farmers Union; John Bosch, national vice-president of the Holiday Association; and Harry Parmenter, Nebraska state president of the Holiday Association and also a member of the Farmers Union. If their code was not accepted, at least for consideration, Reno believed the Farm Holiday would call for a strike. "I see nothing ahead but to strike and fight it out to the finish.... So far, their whole program is a failure and the patience of the people will not continue much longer."[146]

Reno received a telegram from M.H. McIntyre, Assistant Secretary to the President: "Replying your telegram. The President leaving Washington tomorrow and will not return until October fifth. Am referring this matter to Secretary Wallace."[147]

The original reports of the committee — Kennedy, Bosch, and Parmenter — were favorable. Secretary Wallace spent considerable time with the group. Bosch remembers that "there was a great deal of discussion with Wallace in various departments." Wallace seemed to be conceding a minimum price for farm products and President Roosevelt stopped the destruction of food.[148]

He received a letter from John Simpson. "Your letter to President Roosevelt ... [is] very proper and opportune. We must all keep hammering away at him."[149]

In answer to a question from a farmer in Guthrie Center concerning the expenses and results of the three negotiators sent by the Farm Holiday, Reno related that the three had gone in Bosch's car and their expenses were estimated at around $200.00. "We took up a collection and raised $150.00 approximately, but as they are staying in Washington much longer than we expected, their expenses will perhaps be more than was anticipated."[150]

Reno wrote Simpson that the committee had been doing good work in Washington, but his personal opinion was that they would fail to get any real results: "Secretary Wallace's program is evidently to destroy all the smaller farmers who are in distress. Consequently, I see no way to avoid calling a strike, and when this is done, no man knows what the end will be."[151]

On October 10 Kennedy, Bosch and Parmenter sent a letter to Reno. They said that they had met with a number of administrators in Washington. Among these were George N. Peek, administrator of the Agricultural Adjustment Administration; Henry A. Wallace, Secretary of Agriculture; Dr. Fred C. Howe of the Consumers Council; Col. Robert Lee,

Assistant Administrator of the National Industrial Recovery Administration; Henry Morgenthau, Jr., Farm Credit Administration; and President Roosevelt.

President Roosevelt spoke to the group in charming generalities.[152] Governor Morgenthau suggested that any farmer who was threatened with foreclosure or eviction wire him "collect," and he would do everything in his power to prevent farmers from being foreclosed on or being evicted. President Roosevelt agreed to support Morgenthau in attempting to prevent foreclosures and evictions until farm prices rose.

Officials in Washington agreed that the administration had granted to all other industries and groups the right to present to the National Industry Administration codes of fair competition based upon cost of production. They said that there was no legal barrier to agriculture submitting a similar code, and it would receive favorable consideration when presented.

The committee concluded: "We respectfully submit to you our report believing that for the first time in our history official Washington concedes that agriculture is entitled to receive cost of production prices for agricultural commodities as has been conceded and granted to industry."[153]

After receiving the committee's letter, Reno wrote to Iowa's Governor, Clyde Herring. He suggested that Herring call a governor's conference to hear the joint report of the Farmers' Holiday and Farmers Union committee, and to work for united action on the farmers' code and issue a general moratorium on farm foreclosures and chattel sales until the time when farmers obtained cost of production.[154]

Reno also sent a letter to the state presidents and secretaries of the National Farmers Holiday Association. He said that, according to the report of the committee, "while our demands were not acted upon ... they were considered in a friendly spirit, and the right of the farmer to come under the NRA with a code providing cost of production was conceded by the President and the heads of the Department of Agriculture and the NRA."[155]

Reno was further encouraged because both Morgenthau and President Roosevelt said they would do all in their power to stop the dispossessing of farmers of their homes. Reno praised the work of the committee and said they had "really accomplished more than any of us expected."

Then he turned to the question. "It was fully understood in the meeting of September 22nd that if our demands were ignored that a national strike would be called by this committee, and this letter instructs you to ask each county in your state to immediately obtain the sentiment of its members, as to whether a national holding movement should be called at this time, or shall we postpone such a movement until it is fully evidenced as to whether the farmer will be placed on an equality with labor and industry by submitting our code for final approval."[156]

He began to receive letters from Farm Holiday leaders. Bert Salisbury of North Dakota wrote: "The farmers of this county have about given up hope that Roosevelt intends to do anything for them. It has been hard to talk to them all summer, as they were so filled up with false hope. Now they are ready for almost everything. I believe a farm strike could be put over now."[157]

W.R. Hogan of Shelby County, Iowa, wrote that there had been a meeting of governors at Sioux City last year and "it was just kids work what they done so don't be kidded again.... A national strike is our only salvation."[158]

Charlie Peters of Missouri Valley, Iowa, reported: "The ones that attend our meetings are raring to go, as we all feel if something is not done for us before long those of us who are unfortunate enough to live in the dry strip and have lost more than ½ of 3 crops by drought, and last year prices so low they are marked in the red, will be unable to put in another crop."[159]

Governor Langer of North Dakota issued an embargo upon wheat and livestock until production costs were assured the farmer. He also sent Reno a wire asking if Iowa raised spring wheat. Reno replied that Iowa did not raise spring wheat but he would send telegrams to the governors of the Midwest suggesting that they join in an embargo on wheat, corn and livestock. Reno asked Governor Herring if he would follow Langer's example, and he said he would if other Midwestern states also joined in the embargo.[160]

From the members of the Plymouth County Farm Holiday came a telegram: "Give Herring twenty-four hours to declare embargo on livestock and grain in sympathy with North Dakota or call strike at once or Plymouth County will begin strike now."[161]

A Kansas farmer wrote that he was working to organize a Farm Holiday group in his area. He said that he felt helpless when he raised corn and pigs and so on to feed the people and clothe them, and "then they destroy it and reuse acreage while people are going hungry...."[162]

From Mechanicsville, Iowa, came a note saying that the Cedar County Holiday Association had voted unanimously in favor of a strike.[163] A writer from Ainsworth, Iowa, agreed with Reno that "Henry Wallace should be compelled to resign & get out, he is a menace to the farmers interest."[164]

But the Estherville, Iowa, Farm Holiday group voted to "give the government time to see which way the cat will jump."[165]

Reno received telegrams from the governors of South Dakota and

Nebraska that they would not enforce an embargo. Berry of South Dakota said that they had produced very little wheat because of the drought, and Bryan of Nebraska said that both the state and national constitutions prohibited carrying out an embargo. Reno replied to Bryan's wire: "As to constitutionality of an embargo, the constitution, both state and national, have been shot all to hell long ago."[166]

The Farm Holiday Association's officers meeting in St. Paul on October 19, 1933, voted to call a strike. But not all the Farm Holiday Associations went along with it. The Colorado Farm Holiday Association decided not to join in the strike.[167]

Newspapers reported that South Dakota would not strike and that the state president was opposed to a farm produce withholding action. But Emil Loriks wrote Reno that this is "all bunk." The South Dakota president had told the press that he did not want to make any statement as to the plans and details pertaining to the strike until the State Board met.[168]

D.D. Collins sent a wire to Reno: "Confusion reigns relative to strike in Wisconsin.... We are pressed for actual status of strike and must have word direct from you. Papers are filled with propaganda."[169]

From Washington came a telegram from E.E. Kennedy: "Withholding for cost of production program justified necessary compel consideration farmers problem. Believe President shielded from friends exposed to enemies...."[170]

On October 20 President Roosevelt called a cabinet meeting to discuss the farm situation. Reno learned through the newspapers that they had decided to loan the farmers money to buy seed to plant, "in order to raise a crop that will perhaps be destroyed." In discouragement, Reno remarked that he believed the administration's promises were "mere gestures to lull the farmer to sleep that his complete enslavement may be completed."[171]

On October 21 Reno sent out a letter to the state presidents. "By the authority vested in your committee and myself, as national president, I am officially calling the withholding of farm products from the market. A non-buying, non-paying program accompanies this strike order and should be observed as far as humanly possible."[172]

Reno went on to say that in answer to their requests for the cost of production, the Department of Agriculture had given them the "Wallace hog program, which is nothing less than a brazen attempt to bribe the farmer to surrender the little independence he has left."[173]

Following the announcement came some invitations to speak. John Erp wanted him to address the Minnesota Farmers Union. Reno replied that he would be glad to address their state convention. He was also expecting to speak at the Wisconsin convention the following day.[174]

To Hans Sorby, who had written, "Let's put up the last fight we have left," Reno wrote: "I am depending to a large extent upon you boys out in the field staying by me through this movement to gain recognition for the American farmer. The strike this time is fight to the finish. We must stick until we are granted production costs for the products of the farm."[175]

To J.T. Thompson of Iola, Kansas, Reno wrote: "I feel sure that if President Roosevelt fails to bring relief during his term as President that in 1936 you will see a Third Party movement that will sweep this country like wildfire."[176]

Orle Oian of Peterson, Minnesota, wrote that he did not think many farmers knew about the strike. "Whenever I meet farmer friends I generally ask them how the strike is coming over their way, and I have met many that did not know the strike was on. Most of the farmers are not using their radios, and many have no daily paper in order to cut down on expenses. Just today there was one fellow that said, 'I gave them fifty cents last year. What have they done?'"[177]

R.A. Smalley of Detroit Lakes, Minnesota, wrote that he hoped Reno would send him a copy of his speech. He would like to try to get the local papers to print it.[178]

About one thousand farmers of Sauk County, Wisconsin, took a vote on whether or not to enter the strike. The vote stood one hundred percent in favor of the strike. The secretary wrote: "Please keep me in touch with the conditions of the strike as I feel that this must go through or we will be lost forever."[179]

A farmer in Indiana wrote that he read of the strike every day, but believed a good deal of the farm strike news was suppressed by the newspapers. "If the strike were national then your courage might lead toward a new deal to the farmer, but I cannot see the present strike as successful."[180]

In November 1933 the governors of the Midwestern states met in Des Moines and passed resolutions favoring the Farmers Union demands for cost of production of farm products, and a moratorium upon farm mortgages and evictions until cost of production was obtained.[181]

In an editorial in the November 2 *Farm Holiday News*, Milo Reno told his readers that the governors of Midwestern states had presented the Farmers Union demands to Secretary Wallace and been turned down. "This comes as no surprise to me. Secretary Wallace has been opposed to the Farmers Union and the Holiday program and obstructed it in every way possible."[182]

The President also turned down the governors' proposals when they presented them in Washington. Reno wrote a friend, "The Governors were mighty fine in conference here in Des Moines and endorsed our program most heartily, but they sure got slapped in the face in Washington.... Every other industry that has prepared a Code has been conceded production costs as a basis for fixing price. Henry Wallace and President Roosevelt not only refused to entertain the farmers' code, but also definitely and positively refused him the right to cost of production...."[183]

On November 9, 1933, Reno telegrammed a Texas Farm Holiday leader that the strike was on in full force in Iowa, Wisconsin, Minnesota and the Dakotas, and just beginning in Nebraska. However, opponents of the Farm Holiday were working through the county Democratic chairmen and the Farm Bureau to organize antistrikers, "deputizing and arming them" to suppress the activities of the strikers.[184] Reno wrote to a New Mexico supporter that this was a crucial time, "when the plain people of the cities, towns, and farms are fighting with their backs to the wall."[185]

On November 13 Reno wrote in an editorial in the *Iowa Union Farmer* that Wallace, in his speech at Des Moines, had compared the Farm Holiday movement to the "nerve in an aching tooth, you deaden the nerve only as a last resort." Reno observed, "This statement carries a direct threat, that if the Holiday movement does not submit to the dictation of the bureaucracy that is now in control, and rapidly being enlarged, that the movement must be deadened." But, said Reno, if you "deaden the nerve of a tooth, the alarm signal ceases and the tooth may rot and decay without disturbing the patient."[186]

He continued to speak and try to get his message out as much as possible. Through Richard Bosch he made arrangements for D.B. Gurney of Radio Station WNAX at Yankton, South Dakota, to read occasional short statements about the Holiday situation, in his noon-time farmers' program.[187] Reno also gave a speech over radio station WNAX in the middle of November.[188]

Meetings of the Farm Holiday Association were held whenever possible. Fred Plueger, the Secretary of the Plymouth County Farm Holiday, wrote that they held a county meeting and the room was filled to capacity, and "by the expression of the people that attended this meeting the holiday in Plymouth County is not nearly busted as ... [some] say."[189]

In Sydney, Nebraska, Henry Blome reported that the city council refused to let Reno and the Farm Holiday use the city auditorium. The council said they "did not want any radical to stir up the people." Blome said that most of the members of the council were Democrats, and that is where the opposition came from. The Farm Holiday group in western Nebraska could lease the hall for $25. But, "with all the hostilities developing we were afraid to try this as some of us could hardly afford it."[190]

The movement was slowing down and the strike was not effective. In a sense Reno was writing an eulogy for a dying movement when he wrote Orle Oian, the farmer who had complained that he was not hearing much about the Farm Holiday in his part of Minnesota: "The Farmers' Holiday Association has done one thing — it has done more than any other farm organization in existence today to bring about the attention of the entire nation to the plight of the American farmer. It has done more to save farmers from foreclosures and evictions than any other farm organization in the United States...." The Farmers Holiday protested against the Hoover administration and its lack of concern for the individual farmer facing the problems of the Depression, and it also criticized the early New Deal Farm programs under President Roosevelt. But it never did achieve its goal of cost of production for the American farmer.

What did the Farm Holiday movement achieve? In the years 1932–1933 there had been several strikes announced by the national leaders and some scattered local ones called by individuals in smaller areas. There had been several announcements of strikes which were then called off, leaving many of the members confused and uncertain.

They had, as Reno pointed out, brought the nation's attention to the problems of the American farmers. Perhaps they would have been able to do more if money had not been such a problem. But getting out a newspaper, renting a hall or paying for transportation for speakers stretched their limited resources to the breaking point. And lack of money was a symptom of the strained finances of the farmers and also a sign of the lack of strong commitment of many of them. And so the Farmers Holiday movement dwindled and died.

Other groups would arise out of the farms of the Middle West to organize and protest in the succeeding years of the Depression. But none were ever as large as the Farmers Holiday nor ever quite as exciting.

4

"If at First You Don't Succeed": Dan Casement Fights the New Deal

The 1930s were tough times for Kansas farmers and ranchers. They faced not only the low prices of the Depression but also the severe drought and dust storms besieging the Great Plains. One commentator, former Congressman Clifford R. Hope, Sr., wrote that during these years Kansans got a "double dose of misery and calamity."[1]

Among the farmers and ranchers who were facing tough times was Dan Casement of Manhattan, Kansas. Casement owned and operated a 3,500 acre ranch-farm in central Kansas, northeast of Manhattan. He was one of the outstanding cattle feeders of the country, and his steers often won prizes at cattle shows in Chicago, Omaha, Denver, Fort Worth and Los Angeles. He was well educated, a graduate of Princeton University, and a veteran of World War I. In addition, he had written several articles, which had appeared in national journals, concerning his life as a Kansas stock-man and also expressing his independent philosophy.[2]

But in the 1930s he too suffered the problems of low prices and drought. He would later write that his farm "fared badly," and that 1930 was "by far the most disastrous of [his] entire experience."[3]

Like farmers across the nation, Casement watched the political scene with concern. As the 1932 presidential elections loomed ahead, he and the rest of the nation wondered who would be elected and what programs would the candidates bring? President Hoover, the Republican candidate, promised mostly a continuation of the programs of his first administration. But Franklin D. Roosevelt, the Democratic candidate for president, promised new programs. Of particular interest to Kansas farmers were his promises that he would support legislation to alleviate the problems farmers were facing. Departing from their traditional Republicanism, Kansans voted, along with most of the nation, to elect FDR president in 1932. Following his election, and soon after his inauguration, FDR sent an agricultural adjustment bill to Congress. He proclaimed that it was the "most drastic and far-reaching piece of farm legislation ever proposed in time of peace." The bill would cut back agricultural production by providing for reductions in acreage or livestock of certain basic commodities, such as corn, hogs, wheat, potatoes, and cotton. The farmers who made reductions would be compensated by payments from a tax levied on processors.

Mike Sullinger, a farmer near Carson, North Dakota, looks to the skies for rain. An Arthur Rothstein photograph. (Courtesy of Library of Congress, Washington, D.C.)

But Casement, a staunch individualist, refused to join the government's crop reduction program because he did not believe the federal government should regulate the decisions of independent farmers. In addition to his ideological opposition, he also later came to feel that the government farm program was costing him money. Since he did not join the program, he did not receive some of the government payments for reductions. But his hogs probably were sold for lower prices because the processor had to figure in the cost of the processing tax when determining the price to pay for the hogs. In a letter to an acquaintance he wrote, "I must have paid at least $10,000 in processing taxes on the hogs I shipped during the life of the AAA ..." and, "I long ago definitely kissed my hog tax money goodbye."[4]

Thus, Casement was interested when he was contacted by members of the Farmers Independence Council and asked to become president of the organization. It was through leading this group that he would fight the New Deal Farm program.

The Farmers Independence Council of America was formed at a meeting held on April 11, 1935, at the Raleigh Hotel in Washington, D.C. The temporary officers chosen were: First Vice-President, Stanley F. Morse of

South Carolina; Second Vice-President in charge of organization, Walter H. Chappell of Kansas; Secretary-Treasurer, Dr. E.V. Wilcox of the District of Columbia; and members of the Executive Committee, Congressman Fred L. Crawford of Michigan and E.B. Dorsett of Pennsylvania.

They came from varied backgrounds: Morse owned a farm in South Carolina and also worked as a consulting agricultural engineer with various firms. He had been working for the American Liberty League (a group strongly opposed to FDR and the New Deal) for about half a year before the formation of the Council.[5] Chappell had worked in public relations in Kansas. E.V. Wilcox wrote for the *Country Gentleman* and had been formerly connected with the Department of Agriculture.[6] Fred L. Crawford was a Republican Congressman from Michigan, and E.B. Dorsett was a leader in the Pennsylvania Grange.[7]

The group adopted a declaration of principles. They proposed to foster the fundamental right of the individual to independence of speech, thought and action; reestablish the virtues of industry, self-reliance and thrift; and protect the "freedom of every farmer to operate his farm according to his own judgment and to insist that the Government shall not [by any means] ... regiment or attempt to control any farmer in the management of his own farm."[8]

For two months after their founding, the group searched for a President, finally settling on Dan Casement of Manhattan, Kansas. Because Casement had written some articles in national journals criticizing the New Deal, his name had come to their attention.

Stanley Morse, the Vice President of the Farmers Independence Council, sent Casement a letter.[9] The letter was written on paper using the American Liberty League letterhead. According to Morse, his letter had "nothing to do with the Liberty League nor has it in any sense been promoted by the League." His next words contradicted this: "At the same time I have talked to Mr. Shouse about it...." Jouett Shouse was the president of the American Liberty League. Thus, the evidence would indicate that from the very first there was some connection between the Farmers Independence Council and the American Liberty League.

Morse said they had formed an organization and were working on finding financial support for their group. They believed that they would receive adequate funds as soon as their set-up was complete. They had already received some contributions. In the next few days they expected to open an office for their organization, and they hoped to conduct a 16 month campaign.[10]

The next paragraph contains more contradictions. Morse stated: "It is our purpose to make the Farmers' Independence Council a patriotic non-partisan organization. We already have the approval of a number of farm leaders. In addition, we have the unofficial approval of the Chairman of the Republican National Committee and of several of its members, including John Hamilton of Topeka. While the Council will be non-partisan, we have to recognize the fact that if the farmers want to change or abolish the AAA, the only way will be by the ballot, and that must be by voting the Republican ticket."[11]

An interesting part of that letter was the letterhead for the Farmers' Independence Council. The address at the top was 1058 National Press Building, Washington, D.C. This was the same building and same room as the address given by the American Liberty League. Later, Morse wrote that he was glad to have received Casement's wire accepting the presidency of the Farmers Independence Council. He thought that the office would not involve too much personal attention on Casement's part because he (Morse) expected to handle administrative duties and in general direct the operations of the organization. "At the same time there will doubtless be various speeches for you to make, and your council will be most welcome."[12]

Although Morse originally indicated that the duties of President of the Farmers Independence Council would be somewhat limited, Casement soon found himself involved in two major areas: fund-raising and communications.

Casement was asked a number of times to meet with people that might help with financing the work of the Council. On June 21, 1935, Morse wrote to Casement suggesting that he call on George H. Davis at the Board of Trade, Kansas City, and also R.W. Hoffman of Flour Mills of America to ask for help in financing the Farmers' Independence Council of America.[13]

On June 22 Morse wrote that he was still working on the financing, and had two groups who had agreed to make some initial contributions. But at the present there were no funds in the treasury and "I shall have to finance myself."[14]

On June 23 Morse wrote Casement again. He was still concerned about financing, and asked if they were to arrange a luncheon with a few businessmen, could Casement come to Chicago to talk with them at short notice. He also wanted to know when Casement was scheduled to address the Chamber of Commerce. Morse thought he might travel to Minneapolis to see the milling executives there.[15]

Morse wrote on July 6 that William Whitfield Woods, President of the Institute of American Meat Packers, had told him over the phone that he had heard from Casement. Woods said he intended to earnestly request that his group provide some of the financial underwriting of the Farmers Independence Council. Morse said that with Wood's help, and also with that of some important people he expected to see next week, "we may be able to get things lined up pretty quickly."[16]

On July 7 Morse wrote that at a meeting held that evening, at which Woods, Tom Wilson and others were present, $5,000 was pledged for the Farmers Independence Council to become available as soon as $10,000 more was pledged. "I hope to be able to get that and perhaps more in two or three days. The present plan now is to aim at $40,000 initial under-writing." He said that the arrangement he had made with Woods and his group was that Morse would handle all the publicity, as they did not want to appear to be connected with this activity. "In fact, it is OUR organiza-tion, an organization of the farmers and not that of any commercial group, and we must handle it as we think best."[17]

Along the same line, Casement received a letter from Edward Went-worth, Director of Armour's Livestock Bureau. He wrote that the Institute had agreed to make an initial contribution of $5,000 and might give more.[18] The reason so many packers were interested in backing the Farmers Inde-pendence Council was that the New Deal was hurting them financially. Wentworth explained to Casement that the reduction in volume in hogs being brought to market, and the processing taxes on hogs, had seriously hampered all hog processors, especially the small ones.[19]

On September 24 Morse wrote that he hoped Casement could arrange to arrive in Chicago a day or so before his meeting. Morse needed him to go with him to call on a few men. He reminded Casement: "Whether you like it or not, you are a national 'figger.'"[20]

Morse also recounted that he had been after Boylan, President of the Chicago Board of Trade, that day, hoping to find out when their commit-tee was going to act on their contribution: "But [I] could get no definite information.... It is provoking to have things move so slowly when we need to be using our energies in the field."[21]

Morse's next letter was on the same theme: "The main financing is hanging fire, and I am getting pretty much fed up and out of patience be-cause I am unable to do work that should be done to keep things going. For instance, I am unable to get out in the country and see farmers, as I should."[22] On August 31 Morse wrote that financing might be completed next week but he was still financing himself. He wrote, "Am rather sore and fed up with our Chicago friends. Are they playing me for a sucker?"[23]

Morse wrote in October that he had received a letter containing a check from an acquaintance of Casement's. He said it would enable him to pay the office rent.[24]

H. Alexander Smith wrote Casement of some plans for him to come to New York to address the Merchants Association for Greater New York and the New Jersey State Grange. The Grange talk would be arranged by the chairman of the Republican state committee. There would also be a party arranged with some members of the Stock Exchange, a luncheon engagement with the National Republican Club and a meeting in Wilmington with some of the DuPont executives. "I think the point with regard to the financial backing is simply this. [They feel] that if in our publicity it should appear that New York bankers and the DuPont and other companies of that kind are your financial backers, there might be suspicion that this movement had not emanated from the farmers themselves."[25]

On December 23 Morse sent along a letter from Lammot DuPont, who said that he was "entirely in accord with the ideas of the Council, as he understood them having gotten this impression from Captain Casement's remarks and from the literature you [Morse] sent me and am glad to contribute some financial support." He sent a check for $5,000. "Since Captain Casement lunched with Mr. Crane and the others, I have talked with many of those present and they have all expressed themselves as being very much interested and impressed with what he told us."[26]

Within a few days other checks also arrived from the East. S.D. Townsend of the Wilmington Trust Company sent a check and wrote, "I wish it were possible for every voter in the country to hear Capt. Casement. He was most convincing...." J. Thompson Brown of the DuPont Building sent a check for $250. He also was much impressed by Casement's visit to Wilmington.[27]

Thus Casement was very much involved in soliciting funds for the Farmers Independence Council. His efforts and connections provided a considerable portion of the funding the group received. He also made many speeches for the Council, some before large groups of people and others broadcast over the radio. In a letter to Casement, Morse said that they intended to start a radio program featuring Casement. Morse hoped to make arrangements for Casement to give weekly 15 minute talks on the radio.[28]

In the beginning, Casement confessed that his commitment to do a weekly broadcast for six months was causing him "mild mental discomfort." He said that he feared the subscribers to the Farmers Independence Council might be disappointed "should the enterprise fall far short of expectations." And he held a "growing conviction that such might be its fate."[29]

But he received much encouragement from Morse and other members of the Council. Writing on the back of some old Liberty League stationary,

Morse wrote that he had read Casement's second radio speech and liked it. He thought it should appeal to conservative Democrats.[30]

Morse later wrote that the National Broadcasting Company refused to take a contract for their radio programs on the grounds that they did not accept talks on "controversial subjects."[31]

Relenting a little, NBC later called and said they could have 15 minutes free time on WENR, Chicago, which Morse was going to use.[32]

About this time Morse wrote Casement that they were able to get 2:15–2:30 p.m. for Casement to broadcast. He thought the time was pretty poor.

Later they tried making recordings and sending them out free to small local stations. Morse reported that four radio transcriptions of speeches made by Dan Casement were offered to radio stations without charge, and 160 stations had accepted them. But the three biggest radio stations had refused to sell the Farmers Independence Council time. "Proof is ample that the Government is trying to control radio broadcasting."[33]

Casement also wrote letters and articles that were used by the Council. In his speeches and writing Casement brought out the values which he held most strongly: independence and individualism.

On September 21, 1935, Morse wrote that he was giving a report to the American Liberty League and planned to mention the corn-hog program in his report. He said he would welcome Casement's comments as he prepared the statement. Morse suggested the statement might be put in the form of a letter to Wallace, with copies being sent to the papers.[34]

On September 25 the Farmers Independence Council issued an open letter to Secretary of Agriculture Henry A. Wallace. Wallace was holding a corn-hog conference in Washington to consider possible changes to improve the corn-hog program for the future. But the Council suggested that the corn-hog program be ended. The Council authors (primarily Casement and Morse) said they were speaking for the over seventy percent of all corn and hog farmers who had not cooperated in the AAA corn-hog reduction program. Various reasons were given for ending the corn-hog program. One, the corn-hog program had not resulted in increased prices for corn and hogs during 1934 and 1935. Price increases, where they occurred, were due mainly to the 1934 drought. The drought cut the corn crop from an average of 2,600,000,000 bushels to 1,380,000,000 bushels. As to hogs, it was estimated that the drought conditions and feed shortage would have compelled farmers to reduce their hog litters about 25 percent even without the AAA program.[35]

Two, the Council admitted that the farmers who were suffering from drought welcomed the benefit payments received from the processing taxes.

"While these payments were welcome to farmers suffering crop failure from drought, other less costly methods could have been used to relieve such farmers."[36]

Three, the Council opposed the AAA program because it attempted to keep many uneconomical producers operating who should have been allowed to either become subsistence farmers or find other occupations. The Council believed that production should be allowed to return to normal.

About this time Morse wrote an article suggesting that Roosevelt had omitted some things in his speech to the farmers at Fremont, Nebraska. Casement took the idea and enlarged and changed it. His article compared what President Roosevelt had said to a crowd of farmers at Fremont, Nebraska, with the actual situation as he saw it. Roosevelt began by describing the distressing situation in Nebraska when he had visited the state three years before. There were low prices, accumulated surpluses, deteriorating soil and equipment, schools closing, and foreclosures. The situation, he said, was not peculiar to Nebraska but applied to nearly every state in the nation. Roosevelt's description, said Casement, was a "reasonably accurate picture."

Then Roosevelt applauded the farmers for their faith, hope, patience and courage. Casement said these independent self-reliant farmers deserved all the praise Roosevelt could give them. And, Casement added, most of these farmers had refused to sign adjustment contracts.

Roosevelt reminded his listeners of his promise made three years earlier: That he would attempt to meet the situation in every way that human effort and ingenuity made possible; that he would do his best and if unsuccessful, would admit it and try something else. "But," he said, "that was not necessary" because the Agriculture Adjustment Act had been successful. "You know its general results," he said, despite the fact that "there have been many imperfections in it." Casement commented that that was an understatement. "It stinks."

Roosevelt asserted that "for the first time in the history of the nation we have begun to understand that we must harness Nature in accordance with Nature's laws instead of despoiling Nature in violation of her laws."

Casement said that he was not sure what Roosevelt meant. "Surely he doesn't intend to imply that Nature's laws decree the scarcity that his adjustment act has been trying to install. And as for the dust storms, they have been augmented by his benefit payments to wheat farmers who, save for these payments, were already broke and otherwise would have desisted, to their own and the country's advantage, their efforts to farm land that never should have been plowed."

Casement could not accept Roosevelt's evaluation of the "most important gain" coming from the Agricultural Adjustment Act, the development of the farmer's ability, through cooperation with other farmers, to direct and control the conditions of his life. Roosevelt praised the 100,000 committeemen in nearly 5,000 counties who promoted this cooperation among the nation's three million farmers. The government, said Roosevelt, was only the "unifying element" that helped the farmers to cooperate. But Casement commented, "Everybody knows that these numerous committeemen are the most inefficient and least respected farmers in their communities, that they are induced by a daily stipend of public money, that they are in truth a host of hurrying spies, and that the 'unifying element' is in fact nothing more or less than the consumer's tax money."[37]

About this time, August 1935, Casement seems to have had second thoughts about his obligations in the Farmers Independence Council. In reply to a letter from Casement indicating his nervousness about continuing on in the office of President of the Council, Morse wrote, "I promise that you will not be called on for much personal activity, but I shall continue to need your judgment and moral backing. In fact, your withdrawal from the FIC probably would wreck the entire organization, after our weeks of planning and hard work." Morse concluded his letter with the suggestion that Casement go on out to his ranch in the west and "let me do the worrying."[38]

As Casement and Morse and other members of the Farmers Independence Council worked to oppose the AAA, they were encouraged by various court actions. The processing tax of the Agricultural Adjustment Act was being challenged in the circuit courts. Theo. Lampe, officer of a livestock commission company in Kansas City, wrote Casement that he had read in the paper that the Federal Circuit Court of Appeals had ruled the collection of the processing tax was unconstitutional. Lampe believed that the United States Supreme Court would sustain that decision. He continued: "Personally, I feel that a great deal of credit is due you for the firm stand you have taken regarding our constitutional rights...."[39]

Theo. Lampe wrote on August 1 that Judge Otis had granted an injunction to 14 different millers and packers against the AAA tax. "It brought to mind the good fight you have made and the wide publicity you gave the matter."[40]

On January 6, 1936, the Supreme Court announced its decision that the processing tax of the Agricultural Adjustment tax was unconstitutional.[41]

Following the Supreme Court decision, the Farmers Independence Council issued a statement: "The decision of the United States Supreme

Court holding the entire AAA program unconstitutional vindicates our judgment that the AAA was so unsound in every way that it had to be combated until eliminated."[42]

The Agriculture Department decided to call a meeting to discuss alternative plans for a national agricultural program. When Casement requested that he be allowed to attend the conference, Chester Davis, head of the Agriculture Adjustment Administration, replied: "NEITHER THE SOCALLED QUOTE FARMERS INDEPENDENCE COUNCIL OF AMERICA UNQUOTE NOR ANY OTHER AFFILIATES OR REPRESENTATIVES OF THE AMERICAN LIBERTY LEAGUE OR OF THE BIG PROCESSOR GROUPS ARE INVITED TO THE AGRICULTURAL CONFERENCE HERE FRIDAY AND SATURDAY STOP THIS IS A MEETING OF FARMERS AND THEIR REPRESENTATIVES."[43]

Casement sent a telegram to President Roosevelt protesting his exclusion from the Washington agricultural conference: "Chester Davis, administrator of the late AAA, coupled his refusal with the absurd implication that we are affiliated with the American Liberty League or with processors, which indicates complete ignorance or deliberate misstatement."[44]

H. Alexander Smith sent Casement a letter saying that in the *New York Herald Tribune*, January 10, 1936, an editorial appeared entitled "For Yes-Men Only." It quoted from the telegram of Chester Davis to Casement refusing Casement admittance to the conference.[45]

Following their conference in 1936 with selected farm leaders, the Department of Agriculture prepared a new agricultural bill. This bill did not use a processing tax but instead relied on general taxing revenues to provide for payments to farmers who made reductions. The stated purpose of the bill was to reduce production in order to promote conservation. However, many critics felt that its main purpose was to reduce production and raise farm prices in a way which might be acceptable to the Supreme Court.

On March 17, 1936, Dan Casement sent a wire to Henry Wallace: "Press dispatches tonight report your reorganization of administrative branches of old unconstitutional AAA. It is generally understood your purpose under the new substitute law is to retire thirty million acres from commercial production this year. Have you forgotten the Supreme Court ruled control of production to be illegal? You must know the new law's only actual purpose is to give checks to farmers before election. Every day that passes makes the law's ostensible purpose less possible of attainment and the law itself a more hideous joke. stop."[46]

Casement wrote Hoover: "I am surprised and disgusted that no prominent spokesman in the opposition to the new deal has denounced the new farm bill for just what it is. It has no semblance of virtue except in its title, and that is a palpable subterfuge. If, without loud protest, we permit farmers in any considerable numbers to come forward and apply for benefits under this bill, we will simply be playing ball."[47]

Casement noted with disgust that in Kansas the American Farm Bureau was paying most of the expenses of the farmers they sent to Washington to support the AAA.[48]

One writer, who had been a witness before the Senate Committee on Agriculture in opposition to the AAA amendments, said he had observed the farmers who came to Washington "rejoicing at getting paid for doing nothing."[49]

Another wrote: "Have followed all you said in K.C. Star and you are right. AAA Referendum was a joke in this county. Small notice of it placed on inside page of local paper. Scarcely anyone knew of it. Very few voted, but Report was 'seven to one in favor of A.A.A.'"[50]

Soon after the Council organized, the issue of the connection of the Council with the Republican Party came into play. In September of 1935 Morse had written that he had met the assistant treasurer of the Republican National Committee who planned to suggest that the Republican National Committee give its official but confidential sanction to the Farmers Independence Council as the agency to work among the farmers.[51]

But, apparently, there began to be some difference of opinion in the leadership of the Republican Party concerning the Council. In another letter Morse wrote that he agreed with Casement as to the necessity for keeping the Republicans from devising some new programs to be offered to the farmers. Morse said that he had talked to a man closely attached to the party and they were seriously considering such a possibility.

In November of 1935 Morse sent Casement a memorandum concerning the thinking of some of the most prominent leaders in the Republican Party. A friend had written him that "some of the most prominent leaders of the party in the Middle West keep thinking that they must either be silent on AAA or else offer the farmers what they call a substitute."[52]

In January of 1936 Smith wrote that he was doing all in his power to keep party leaders from writing a farm plank in the Republican platform that differed from Casement's position.[53]

At the same time, Kurt Grunwald received a letter from William Jardine, former Secretary of Agriculture. Jardine said he thought the Republicans would "commit political suicide if they don't have a plank in their platform that is designed to keep agriculture from slipping back to where it was before Roosevelt came into power.... The farmers are feeling more prosperous. Roosevelt is going to get the credit for it whether he deserves it or not. …I don't think the triple A is sound; I think it is headed for the rocks. Nevertheless, these benefit payments which have been so generously distributed to the farmers are being cordially received, and the recipients, I am sure, are not going to forget their Santa Claus when next he is in trouble and needs their vote unless someone else as big or as good comes along in the meantime."[54]

Morse and Smith also wrote former president Herbert Hoover at this time, sending along one of Casement's speeches: "It occurred to me that you would be interested in reading his hard fisted, head-on attack on the A.A.A." Hoover replied, "My dear Alex — I understand the DuPonts are supporting the group you mention. Casement is making a good fight."[55]

The former president stopped to see Morse while he was in Chicago. Morse was not in the office, but Hoover left him a message. "I have been endeavoring to lay some foundations that might be helpful to your work."[56]

Morse replied to Hoover a few days later, promising to send an address that Casement was scheduled to present, and commenting: "The key to the elimination of the New Deal is in the farmer vote."[57]

Early in the campaign of the Council, the Department of Agriculture began to respond to the Council's criticisms with attacks of their own. Morse sent along a letter from Charles Burkett, who had been working for the Farmers Independence Council in the East.

Burkett wrote: "One of these days you may not have a vice-president. This morning I had a caller — a high official in the U.S. Department of Agriculture, who said he had been asked by an AAA official, both men knowing me, to be careful what I say about the AAA. The caller didn't think that I, an old agricultural worker, should have opened up in the manner I had — that those articles were doing lots of damage to their work.… I was told they were looking up my address to get in touch with me (or after me)."[58]

Yet some support was also surfacing. Morse also said that he had been in Iowa with Jim Howard of the Iowa Farm Bureau for three days and found conditions "very favorable for starting our work."[59]

He found Howard to be "most helpful." Morse believed that considerably more than 50 percent of the farmers were not sold on the AAA: "Many of them are beginning to resent domination by county agents, coercion by their own committeemen and centralization of authority in Washington. They fear, especially the younger ones, the threat of an increasing enormous tax burden."[60]

Morse said that L.J. Tabor of the National Grange was also critical of the New Deal's agricultural program. Morse wrote, "We now have 400 members in 39 states. Over 20 memberships received today."[61]

Morse reminded Casement that he was scheduled to speak at the Executives Club at the Sherman Hotel in Chicago, as well as make a trip to Indiana and Ohio. "In the last three days we have received additional contributions of $1700, so I guess we can manage to pay your expenses."[62]

Then on February 10,1936, new troubles appeared. Morse sent a letter to the directors of the Farmers Independence Council. "As you have seen by the newspapers, the Black Investigating Committee of the Senate is commencing to investigate certain patriotic organizations such as the Liberty League and the Crusaders.... While the Farmers' Independence Council has not been included in this investigation yet, it probably will be."[63]

On April 7 Morse sent a short note to Casement. He said that the promised financial support had not come in, and he was thinking about liquidating at the end of the month. "Am very worn down with work, worry and fight, and cannot see very clearly just what steps to take as I seem to be little good for raising money. However, will try to hang on, but will not exceed our funds. Any suggestions?"[64]

On April 9 Morse wrote Casement that he had received a subpoena from Senator Hugo Black and the Senate Lobbying Investigating Committee. He had decided to appear before the Committee because the lawyer for the Farmers Independence Council had advised that he comply with the Senate committee's request. Morse also wrote Hoover, "I have just received a pressing invitation from the Black Committee."[65]

Hoover replied, "I hope you land into the Black Committee with all your might."[66]

Casement sent off a telegram to Black protesting the committee's actions. Casement said that the committee was prying into the affairs of an organization that was not formed for lobbying and had done no lobbying, but simply because the group was "effectively opposing unamerican, uneconomic policies of the present administration."[67]

The Senate Special Committee to Investigate Lobbying Activities questioned Morse for three hours. Morse said that he was a consulting agriculturist who had worked for different groups making reports on land evaluation, timber assessment, economic surveys, etc. He had also worked for the American Liberty League reporting on the AAA.[68]

Following his appearance before the Senate Committee, Morse wrote Casement about his evaluation of the experience. He said that the purpose of the Black investigation was to discredit and destroy the Council because it opposed the Administration's attempts to control the nation through its appeal to farmers. He said that the investigation tried to show that the Council was connected with the American Liberty League, and that the Council was only a smoke screen for "big business."

Morse's defense was that he needed the aid of large contributors, such as S.M. Swenson of Swenson-Texas Corporation, the Chairman of the Board and Director of the National City Bank, and Lammot DuPont of the Delaware DuPonts. He felt that the Council could not have functioned without the aid of large contributors because "most farmers are not sufficiently informed and alarmed to see its need."

And it is probably true that the Farmers Independence Council was able to secure more publicity because of outside financial help. But, of course, Morse was trying to hide the downside of this, which is that the Council represented not only the farmers but also its financial backers, the businessmen who resented Roosevelt and wanted to work for the election of someone else.

Also, the committee tried to show that the Council was a political organization connected with the Republican Party, but Morse said that the Council had criticized the agricultural policy of the Republicans also.[69]

In a letter to members of the Farmers Independence Council, Morse related that the Black Committee had assumed unconstitutional powers when they had examined letters written to him in confidence. "A telegram sent to our secretary, Dr. Wilcox, on the 10th apparently was in the hands of [the] Committee a few hours later."[70]

Kurt Grunwald, a field worker of the Farmers Independence Council, was questioned by the Committee but refused to give names of the persons he talked with in the various states he visited. He said, "I am not going to get anybody in trouble because I give the names."[71]

When Morse wrote his report to the members of the Council he related with pride that the Committee did not get the list of members of the Farmers Independence Council and only a little of the correspondence of the group. He also pointed out that the Black Committee had not investigated any pro–New Deal group, only those who opposed the Administration.[72]

Shouse, president of the American Liberty League, issued a statement: "Whatever may be the attempt of the Black Committee to connect the American Liberty League with the Farmers Independence Council, I assert definitely that there never has been any connection between the two organizations: The league had no hand in the organization of the council, no part in the financing [of it]."[73]

This may have been technically so. But all along there had been many ties between the two organizations. Morse had worked for the Liberty League before he worked to found the Farmers Independence Council. Other members of the Council Executive Board were also members of the League. The groups shared offices and stationery for a time. Many of the contributors to the Council were also contributors to the League. They liked to claim that they were separate, but evidence suggests that they may have been strongly connected.

Following the Black Committee investigation, the leaders of the Farmers Independence Council viewed the approaching Republican nominating convention with mixed emotions. Morse had written to Hoover that he had studied the candidates for the Republican nomination for president and was "much discouraged with the outlook." He had come to the conclusion that the "welfare of the country" demanded that Hoover be the next president and suggested that he announce his candidacy for the nomination. Hoover wrote back: "That was a kindly and encouraging note."[74]

It appears that Hoover liked to think that there might be some support for a second term for him. He probably wondered if such support might show up at the Republican Convention. At least he continued to show his interest in the activities of the Council.

When Landon won the Republican nomination, Casement sent him a telegram of congratulations but criticized some of his promises to farmers. "It shall be my aim henceforth to prevent these unfortunate pronouncements from deterring you in your tremendous task of saving our American democracy, and to that end I pledge you my hearty support."[75]

Morse, who had been at the Republican National Convention, headed back for Chicago, writing Casement that he believed the basic issue had not changed at all. They must eliminate Roosevelt and the New Deal or be prepared to resist "a determined attempt to establish communism in America."[76]

Apparently, Casement was again feeling discouraged and wished to resign his position as president of the Farmers Independence Council. Morse wrote him that it was their duty to try to arouse the farmers of America to their perils under the New Deal's agricultural program. He believed that there were millions of Americans who would respond to their leadership. He concluded, "We are all depending on you to help us carry on until November. Please don't desert us."[77]

A few days later Morse wrote Casement that he hoped he would not "communicate your rather pessimistic attitude to the other members of our group.... Dan, your leadership has been an inspiration to all of us and we need you."[78]

Morse was also feeling somewhat discouraged. He was concerned that leaders in the Republican Party were not giving the Farmers Independence Council the support they had expected. He visited the committee members and wrote letters to Harrison E. Spangler, a member of the committee, and John Hamilton, chairman of the committee. The point of his remarks was that the Council, if funded by the Republican Party, was ready to go to work in the Middle West with a "corps of experienced speakers and workers." He concluded, "Please do not get the impression that we are primarily interested in fostering the Farmers Independence Council as an organization. We are interested in it mainly as an instrumentality which it is believed can do more effective work among the farmers than can Republican political workers. We have already spent over a year and $27,000 laying the foundation for this work, and this machinery is now available to your organization."[79]

Morse wrote several letters to Hoover on the matter. Hoover replied that he hoped the National Committee would "keep your organization alive and in the field." He promised to discuss the matter with some of the party leaders.[80]

Morse wrote Hoover: "As you know, our organization is the one and only farm group that has had the courage to come out against the New Deal.... They tried to destroy us by various means, including the Black Inquisitorial Committee, but we have survived.... For many months now we have operated on a hand to mouth basis, believing that when the nominations were over we would receive necessary help from the National Committee."[81]

Morse wrote to Crane at the DuPont building that the Farmers Independence Council was suffering from a scarcity of funds and having to mark time until the Republican headquarters approved their activities for contributions.[82]

He also wrote to Arthur Curtis of the Republican National Committee, again asking that the Republican Party finance expanded operations of the group for the 1936 political campaign. He sent a copy of the letter to Hoover.[83]

Hoover replied that he had "no weight" with the present national committee and was sorry to see they were "not cooperating."[84]

The Senate Committee investigations may have hurt the Farmers Independence Council. Grunwald wrote Casement that he had heard the Republican National Committee wanted to eliminate or sidetrack the Council because it had been branded as part of the American Liberty League.[85]

Even if they wanted to avoid allying themselves with the Council, Republican party leaders found the expertise of the Council members useful. The Republican National committee hired several of the members of the Farmers Independence Council to work for them. Grunwald went to Colorado under Republican direction, and Burkett was appointed director of agriculture for the eastern branch of the Republican National Committee.[86]

Despite their declining funds, the leaders in the Farmers Independence

Council continued to speak out in the months preceding the November election. Morse wrote Hoover that they had decided to go ahead with their plans without attempting "any special cooperation with the National Committee." Hoover wrote Morse that he was "delighted they were in motion again."[87]

It is difficult to know what effect the Farmers Independence Council efforts had on the election. Some members of the Democratic Party used the Farmers Independence Council to frighten farmers away from supporting the Republican Party. In October Secretary of Agriculture Wallace, speaking before an Iowa audience, said that the same old crowd were "behind the throne" of the Republican Party. They were the big packers, the Wall Street investors, and captains of industry and finance who were "wearing false whiskers and smoked glasses." He said that they called themselves the American Liberty League, and that the agricultural branch of the League was called the Farmers Independence Council.[88]

Following Roosevelt's overwhelming victory at the polls in November of 1936, Morse wrote Casement: "You were right in your prediction and in your size up of the degeneration of the majority of U.S. citizens. I couldn't believe our countrymen had deteriorated so and I still find it hard to comprehend. The White House S.O.B. had it figured out right and took advantage of the people's weakness and distress to bamboozle and bribe them."[89]

It looked as though the work of the Council was finished. Morse began to close the office. The office secretary found a new position. But new hope returned to the organization when a letter came from Lammot DuPont saying he thought the Council should continue its work.[90]

Financing still continued to be a problem; the group operated on a shoestring. A preliminary financial statement of the Council shows that the total receipts of the group for 1936 were $46,794.20. Of this, $10,000 was spent on printing and distribution of literature, $10,000 on organization expenses, $4,000 on publicity, $2,000 on radio transcriptions and $16,000 on salaries. Telephone, postage, office rent, stationary, etc. took up the rest.[91]

In the following years, Morse was active in campaigns to arouse people to oppose Roosevelt's proposal to increase the size of the Supreme Court, and also worked to oppose the executive reorganization bill.[92]

Casement tried, with little success, to secure the publication of articles he had written on agricultural problems.[93]

He also discussed the possibility of forming a new party with various leaders in the Republican Party. Most of them told him that the idea had also occurred to them, but they felt that it was not practical. They believed that the conservatives in each party should work to regain control over their own parties and defeat the New Deal in that way.[94]

Casement continued to voice his opposition to the New Deal farm program. In December of 1937 Morse wrote Casement that perhaps Casement's vigorous protest at a regional hearing on agricultural programs held by Senator McGill in Topeka, Kansas, had an effect. Senator Frazier had complained in a speech before the Senate that he felt 75 percent of those present at the hearings were handpicked government supporters.[95]

Small signs of farmer discontent with the AAA began to appear. Morse wrote that he was encouraged by the Grange's opposition to parts of the new farm legislation.[96]

In February Casement received an invitation from the president of the Riley County Farmers Union, a branch of the Farmers Union that met near Manhattan, Kansas, where Casement lived. Hawkinson asked Casement to speak at their annual meeting. He wrote that the board decided to ask Casement because they wanted their speaker to be someone who "does his own thinking and is nobody's rubber stamp."[97]

Casement began to speak at local farmers and ranchers meetings. Casement also received a letter from Morse saying that he was "hoping to get back into the fight actively pretty soon. Nothing definite yet, but fighters surely are needed."[98]

The Farmers Independence Council certainly had its ups and downs. It was never very large, having only 400 members, and did not affect the 1936 elections in any perceptible way. But they may have had more of an impact than first appeared.

In April of 1938 Morse wrote to Casement in great excitement that a new movement had started in Macomb County, Illinois. It was the Corn Belt Liberty League. Morse was much encouraged to learn that one of the organizers, G.C. James of Good Hope, Illinois, secretary of the organization, had been a member of the Farmers Independence Council. Morse triumphantly concluded, "Here is direct evidence that the seed sown by us is bearing good fruit. This organization may spread over the entire Corn Belt. Our efforts were not lost."[99]

Casement's criticisms of the New Deal Agricultural Program are different from those of Hirth, Simpson, and Reno. They criticized the New Deal because it did not go far enough to protect the little producer. But Casement criticized the New Deal programs because he felt they went to far, and interfered with the freedom and independence of every farmer. The reaction of the Democratic administration to these groups seems to have been similar, however: Silence the critic. With Reno and Simpson promises were made but not kept. With Casement and his group, a Congressional investigation was lead by Senator Black, who later was appointed a justice of the Supreme Court.

5

"We Want Our Money Back": The National Farmers Process Tax Recovery Association

The National Farmers Process Tax Recovery Association began in Iowa, and many of the group's leaders had been active in the Iowa Farmers Union.[1] They were a strange mixture of radicals and conservatives, united by their common agricultural background and a distaste for the New Deal. The group organized in 1936 and fell apart in 1941. In the words of one of the group's early members, it became a "long drawn fight."[2]

They were drawn together in the 1930s and early 1940s, facing the problems farmers faced at that time, to fight for the recovery of the processing tax, which had been an essential part of the Agricultural Adjustment Act. They believed that farmers had paid the tax and they wanted a return of their money when the tax was declared unconstitutional.

By the terms of the act, processors had paid a tax on the processing of a number of agricultural goods, and this money was then used to pay farmers for cutting their production of these goods. Corn and hogs were among the items on which a processing tax was levied under the Agricultural Adjustment Act.[3]

When the act was passed, farmers greeted it with enthusiasm. But for many farmers, enthusiasm soon waned. Some said that the large farmers were getting an unfair advantage through the AAA program. C.P. Rusch, Secretary-Treasurer of the Iowa Farmers Union, commented that he had talked with "quite a number of farmers in regard to the method which is being followed to bring about a decrease in the number of hogs produced.... To me, as well as to others, it seems very unfair that the Administration should follow a method whereby the farmer who is most to

Fred Maschman feeding his 15 hogs, one of his major sources of income. (Courtesy of Library of Congress, Washington, D.C.)

blame for the surplus should now reap the greatest benefit by the Administration's program."[4]

Many complained that the processing tax was not really on the processors but instead charged to the producers. One farmer wrote, "I am a hog grower and protesting the present processing tax which is in fact not such a tax but rather a tax on the grower."[5] Another farmer, from Clarinda, wrote: "Your theory is that the processing tax is or should be passed on to the consumer. If the price of meat could so easily be raised, do you believe that the packers would have carried such large stocks of meat over into the packing season?"[6]

One of the strongest critics in Iowa of the Agricultural Adjustment Act and the hog processing tax was Milo Reno. Following the passage of the Agricultural Adjustment Act, Reno hesitated, reluctant to oppose the program which he hoped would help the farmers. But after the creation of the National Corn-Hog Committee, which met in Des Moines, Iowa, Reno began to opposed the New Deal. He disapproved of the people chosen to implement the program. The leaders chosen by the Department of Agriculture were well-to-do farmers, county agents, members of the Farm Bureau and university professors, not the "common dirt farmers" who were members of the Farmers Union or Farm Holiday.[7]

As president of the Iowa Farmers Union, Reno had long fought the Farm Bureau in Iowa. Reno also opposed the county agents, men and

Farmer and county agent. (Courtesy of Library of Congress, Washington, D.C.)

women who worked with the farmers in agricultural extension but who had special ties with the Farm Bureau.[8] Reno opposed the Farm Bureau and the county agents for some of the same reasons that he opposed the AAA — because he felt that they took away the farmers' freedom. He believed that cost of production would have brought higher prices without government regulation. Although many Farmers Union leaders had argued in 1933 for cost of production legislation, Congress had enacted the production controls measures of the AAA instead. Reno compared the AAA to a "highway robber" because it did not give farmers a free choice; they either had to "sign this contract [and] deliver," or the government "will take $2.25 upon every hundred pounds of pork you produce [and] ... will boycott and harass you in every way possible." [9]

Reno did not publicly censure President Roosevelt, who was quite popular nationwide.[10] He continued, however, to criticize the New Deal and especially the programs of the Department of Agriculture and its Secretary of Agriculture, Henry Wallace. He made numerous speeches expressing his opinions. In May of 1934, speaking before a crowd of 2000 members of the Farm Holiday Association, Reno called Wallace "the worst enemy the farmer has ever had in an official position."[11] Speaking over radio station WHO on Sunday, March 17, 1935, Reno said, "The things that Henry Wallace stood for before he became Secretary of Agriculture, he repudiated in toto afterwards. Certainly he was the leading spirit in fixing the policies and the program of the Agricultural Adjustment Act. He and his associates conceived the idea of an 'economy of scarcity,' the exact opposite of an 'economy of plenty.' Under his direction, millions of hogs were murdered; under his direction, productive land was taken out of cultivation."[12]

Reno continued to attack the Department of Agriculture and the AAA through radio speeches, editorials in the Iowa Farmers Union paper, and around-the-state "stump speeches." He urged farmers not to join the AAA. Perhaps because of the stand he and other leaders in the Farmers Union took, many farmers throughout the Middle West refused to sign contracts with the AAA corn-hog program.[13] In some states, one-fourth to a half of the farmers refused to join. In Iowa in 1935, out of 140,000 farmers producing hogs, approximately 40,000 did not sign the AAA corn-hog contracts.[14]

Following the Supreme Court decision in January of 1936 declaring the Agricultural Adjustment Act unconstitutional, the Department of Agriculture invited some farm leaders to Washington to confer on proposals for new farm legislation. Although not invited, Reno went to the conference but was disappointed with the results, as the Department of Agriculture proposed legislation providing for production control as part of a program of conservation.

Returning to Iowa, Reno encouraged farmers, particularly those who had not signed the AAA crop reduction contracts, to join together to get their money back. Reno suggested farmers should meet and organize, so that together they could work for the recovery of the hog processing taxes, writing to one farmer, "Certainly the farmer who did not sign and who gets no processing tax back has a right to recover. I think the first step should be to interest all farmers in each county, who did not sign the contract, and establish a state organization to raise the funds necessary to pay expenses and, if necessary, send a committee to appear before Congress with their demands."[15]

Reno contacted various friends around Iowa, some members of the Farmers Union and some not, suggesting that they form an organization for the purpose of claiming the processing tax. One of the friends that Reno contacted and asked to become president of the organization was Donald Van Vleet of Greenfield, Iowa. Van Vleet remembered years later that Reno had called him up after the Supreme Court decision to discuss his idea of a new organization.[16]

Reno's proposal to form an organization was carried out by some of his friends and supporters. On March 12, 1936, twenty-seven Iowa farmers met at the Farmers Union building in Des Moines to discuss ways of getting their money back from the processing taxes that had been levied under the Agricultural Adjustment Act. They believed that the processors had deducted the process taxes from the payments they made to hog farmers for their hogs. They said, "We want our money back ..." and organized to achieve their goal.[17]

Donald Van Vleet was elected temporary chairman. In order to establish his credentials as a leader, and to further encourage the formation of the group, Van Vleet told of his previous experiences with opponents of the processing tax on hogs. Van Vleet said that there had been a meeting at the International Livestock Show in Chicago of those opposed to the processing tax, and such organizations as the anti–New Deal American Liberty League were represented. At this meeting he had been designated to contact different packers in the country to find out what their attitude was on the processing tax. He said when he visited the different packer plants in the country the officials all admitted that they had not paid the processing tax.[18]

A.J. Johnson, a farmer from western Iowa who had been so active in Reno's Farm Holiday movement that he was jailed for several days for his activities, led part of the meeting. This dignified middle-aged farmer, Sunday School teacher and father of seven children expressed the hope that both the farmers who refused to sign the corn-hog contracts and those who did sign them would be able to get their processing taxes back. Even if they did not recover their taxes, he noted, the formation of the group would be a "hindrance" to any attempt in the future to "put out any other such measure as the last Triple A."[19, 20] Reno was not present at the first organizing meeting of the group because he was sick and had gone to Excelsior Springs, Missouri, to recover (he later died there), but his advice and spirit still guided the organization.

The question was also raised at the meeting as to how many hog producers in Iowa had not signed the AAA corn-hog contract. The answer given was 40,000 in 1935 and somewhat less in 1934. Illinois had about 80,000 non-signers.[21]

The group set a membership fee of two dollars, and 22 paid their dues as charter members. They came from all over Iowa; their addresses are of rural routes and little towns scattered throughout the state.[22] Some of them were members of the Farmers Union, some had been members of the Farmers Holiday Movement and some were members of the Farm Labor party.[23] At the close of the meeting they sent a letter to Reno thanking him for his efforts in "laying the foundation of this organization" and saying that they were hoping for his speedy recovery so that they might again enjoy his fellowship.[24]

Following the meeting, cards and letters were sent out to prospective members, and support continued to increase for the cause of recovering the processing tax. One farm woman wrote: "Our hogs died in 1933 so we're out of luck, no use in signing AAA, nor could we sell our corn, which we gladly would have done but had to sell cheap. Thousands of farmers in the same fix as we, sure hit us hard, big fellows had the benefit. First it is hard to get the non-signers together, then it is a long drawn fight to get the Pross. Tax back, if ever. But the AAA did not play fair. Glad the Sup. Judges decided it the way they did. Hope the Farmers will have more to say in the new AAA. Gentleman book farmers make the rules or law!"[25]

Other farmers often echoed her complaint that the AAA benefited the "big fellow" and the small farmers suffered under the program.

Another farmer wrote: "Have been a Holiday Member, one of the first, when organized here and never signed up for the AAA. And if there is any chance in getting my processing tax there is quiet [sic] a few around here that are non signers. I wrote to Mr. Reno and he stated they would organize in each state. I see by the Holiday News you are for Iowa. Would like to get what I think is coming and [learn] how to go at it to get it."[26]

As the Farmers Process Tax Recovery Association grew in Iowa, its leaders attempted to form similar organization in other states of the Midwest. Many of the people they contacted had been active in either the Farmers Union or the Farmers Holiday. Such people as Fred Winterroth of Illinois, John Erp of Minnesota, Harry Parmenter of Nebraska and Emil Loriks of South Dakota were contacted. John Erp of Minnesota eventually became the head of a processing tax recovery organization there. Harry Parmenter served on the board of the process tax association when it became national, and Fred Winterroth worked in the processing tax recovery fight in Illinois.[27]

In South Dakota D.B. Gurney, president of Gurney's Seed Company, endeavored to help the farmers recover the processing tax. He used his popular noon-hour farm program over radio station WNAX to urge his listeners to collect their hog receipts, send them in to him, and enroll in his program to recover the hog processing tax.

Other organizations for the recovery of the hog processing tax were established in Minnesota, South Dakota, Illinois, Nebraska, Kansas and Missouri.

Soon after the Iowa Farmers Process Tax Recovery Association was formed, Donald Van Vleet, the president of the organization, received a letter from C.O. Dayton, a tax consultant in eastern Iowa, offering his services in securing the tax refund. Dayton then enrolled many farmers in his area in the program. In 1936 the Iowa Farmers Process Tax Recovery Association and Dayton filed about 265 claims with the commissioner of Internal Revenue. The claims were rejected, however, on the grounds that the farmers did not have receipts showing they had paid the tax. As A.J. Johnson commented, "They were all turned down because we couldn't prove we were taxpayers, everyone knew we bore the burden."[28]

During this time, leaders of the struggling group looked around for financial help. In June 1936 Van Fleet wrote to a member in Illinois: "I don't suppose that you received very favorable reports from the Liberty League either."[29] Later he was to contact several wealthy conservative owners of large farming establishments with requests for aid.[30]

Van Vleet also visited some of the packing companies in Chicago, searching for support. He visited Swift and Company in December 1937 and later urged them to support those "that have proven their friendship and are doing a tremendous job in trying to keep the farmer from being regimented. As we know, if this takes place, it also throws industry into the same position. Another process tax on hogs will just ruin the meat industry and the farmer because I doubt if we have a Supreme Court now that will declare is unconstitutional."[31] There is no evidence to indicate, however, that the packers provided financial help to the Recovery Association.

After their failure with the Commissioner of Internal Revenue, the officers of the Recovery Association talked with a lawyer, Frederick Free of Sioux City, to see if he would represent the Association in a suit against the government. He agreed to represent the group.[32] George De Bar, a director of the Association, wrote, "We have assumed a responsibility to our members so I do not feel we should take our licking layed down."[33] Eventually, Free decided that a suit in the courts had little likelihood of success, because it was impossible to file all the cases under one head and each case should be filed separately. Therefore he recommended legislation.[34]

Donald Van Vleet wrote several congressmen asking their help in getting a hearing on a bill for the recovery of the hog processing tax. He wrote to Representative Hubert Utterback (D-Iowa): "Don't you feel that those men who had the stamina and intestinal fortitude to stay out of the AAA at great personal loss of money should be compensated for this?... We are sending you full information regarding our association and hope you will study it carefully and ... use your influence in the United States Congress to give us a hearing."[35] Utterback did not reply. Clifford Hope (R-Kansas) replied that he did not have time to help the group.[36] Republican

Congressman William Lemke of North Dakota felt that it would be difficult to find all the farmers who paid the tax.[37] Democratic Senator Guy Gillette of Iowa believed that some day in the future someone would sponsor a bill to give the non-signer his money back, but that was as far as he would go.[38]

Van Vleet also wrote Dan Casement of the Farmers Independence Council concerning support for the effort to recover the hog processing tax. Casement wrote back: "My judgment tells me that such an effort would be quite hopeless…. There is no conceivable way of determining exactly how the tax was paid or of returning it to producers. I am satisfied that a suit undertaken for that purpose would end in nothing. In this belief I long ago definitely kissed my hog money good bye."[39]

Still Van Vleet was not discouraged. He wrote to John Erp of the Minnesota Farmers Union and a director of the Recovery Association: "I think this battle is going to eventually narrow down to a legislative one, and if we could just line up a few progressive Representatives to fight this thing out in Congress, I actually believe that the conservatives would join in line with them to see if the man who objected to the AAA program and did not cooperate was compensated."[40]

Help came from another source. In June of 1937 A.J. Johnson, the secretary of the Recovery Association, received a letter from Edward E. Kennedy. Kennedy had read an article in the *Iowa Union Farmer* on the recovery action the Iowa hog raisers were planning to get their money back. He had also received a letter from Christian Grell, a member of the Association, asking him to get a hearing on the subject before a Congressional committee. Kennedy wrote, "Of course, A.J., I am at the service of these farmers, but I cannot act without official authority from the association."[41] In effect, Kennedy was asking them to employ him as their lobbyist.

Johnson passed the letter from Kennedy on to Van Vleet, and the two agreed that Kennedy's help might be useful in securing passage of their bill. A group of the Iowa Farmers Process Tax Recovery Association leaders went to Washington to confer with Kennedy.

Kennedy had been active in the Farm Holiday movement. He was a friend of Reno and had led in a memorial service for Reno in Washington, D.C., at the time of his death.[42] He was a member of the Iowa Farmers Union, and in later years was to boast that he was one of the first to propose the penny sales used by farmers to prevent the foreclosure of their property by auction during the Depression. Kennedy served as National Farmers Union secretary and editor of the *National Farmers Union* paper from 1932 to 1936, when he was ousted from that position in a close-fought battle.[43] Since that time he had been lobbying for various state Farmers Unions and publishing a newsletter from Washington.[44]

The group decided to accept Kennedy's offer. Van Vleet wrote Kennedy: "If we could only get a hearing on this and present our side to this case there are things that would come out that I am sure would compel the passing of legislation favorable to us.... If you can secure for us a hearing on this, we will drive to Washington with statistics, affidavits, and evidence that will prove beyond any contradiction that we paid this processing tax."[45]

A.J. Johnson and Van Vleet, with a few others, traveled to Washington to meet with Kennedy and also with various congressmen. Kennedy took the Farmers Process Tax Recovery Association leaders to meet members of the Senate and House Agriculture committees, including Senator Guy M. Gillette and Congressman Fred C. Gilchrist, both from Iowa. Johnson later reported that they must have met with 20 or 25 members, "and they all ... advised legislation."[46]

One of the Congressmen, Sam Massingale of Oklahoma, was particularly enthusiastic about the bill and said, "I will vote for this bill and I am damned sure it will pass." In commenting on the trip, Van Vleet said that they did not meet any opposition, and even processing tax officials admitted the "justice of our claims, although they will not give us the permission to quote their ... belief."[47]

The congressmen suggested that William Lemke and Kennedy work on drafting legislation. The two had worked together before on other projects.[48]

Despite administration opposition, Lemke had been able to achieve the passage of a farm mortgage bill through Congress. Then it was sent on to President Roosevelt who signed it. Representative Lemke had accomplished a major legislative feat. The farm leaders hoped that he would be able to achieve a similar feat in securing the passage of legislation returning the processing tax to the farmers.

In early August Kennedy sent a telegram to Donald Van Vleet and the members of the Farmers Process Tax Recovery Association stating that he had secured introduction of House Resolution 474 by Congressman Lemke providing for hog producers securing refunds of the processing tax, and was arranging for introduction of the resolution into the Senate. He also commented that he was confident that they would be "pleased with the simplicity of refund procedure we have incorporated in measure."[49] Donald Van Vleet relayed the news to Recovery Association member Paul Bock: "The old reliable firm of Frazier and Lemke, which has passed so much constructive farm legislation, is now back in the fight and is going to do everything in its power to give justice to the non-signer."[50]

Lemke was interested in the progress of the Recovery Association.

He wrote Kennedy in October, saying that he was wondering "how you are getting along with the Processing Tax. I find that the farmers are in real need of a fighting organization ... [because] the various farm organizations are rather representing the Department of Agriculture than the farmers."[51] Kennedy wrote back that he had been busy that last seven weeks in setting up the framework and getting the manpower in the Farmers Process Tax Recovery Association organized.[52]

Kennedy also made a significant addition to the group when he found a report put out by a bureau of the Department of Agriculture which would support their claims that the farmers rightfully deserved a processing tax refund. Kennedy wrote to Van Vleet in great excitement in November of 1937, saying that he was sending a document that had been prepared by the Bureau of Agricultural Economics. This document concluded that the processing tax was borne by the hog producer and confirmed that packers had made statements to the same effect. Kennedy also had prepared a statement that could be sent to the county newspapers to get the information out to the farmer. And then, most important, Kennedy said, was to "keep up the drive for members—weekly letters to the boys, to all of them — it will take members, thousands of them before we can make a mark here on this bill. I have talked to many additional members of Congress willing to support the bill — it's just, they agree — but does the farmer want it? If he does he'll get it, that's the tone — that is what it takes."[53]

The report, Kennedy exclaimed, "proves the general fact conclusively that the processing tax on hogs was borne entirely by the farmer and not the producer, distributor, retailer or consumer. This is what we wanted to know officially, wasn't it?" He said that he was sending copies of this "extraordinary document," marked for their convenience. He did not mention that the writers of the Bureau report believed that farmers who signed contracts had been repaid for their losses by the federal payments. He used only the parts of the report that bolstered his arguments.[54]

He suggested that the Association leaders confine their use of it to the effects of the processing tax on hogs. "That is what we are interested in NOW, isn't it?" He was sure that the facts of the document would inspire the state leaders and get them to redouble their efforts and "fire all your men with enthusiasm to continue their activities as rapidly as possible."[55]

But Kennedy realized that arguments alone could not produce legislation. He urged the leaders to recruit more members. The Association needed thousands of members with thousands of claims to get a bill passed. At the time he felt, "there is no opposition, in fact the reverse is true, but they [Congress] won't fight for it until they are convinced the farmer who has the money coming wants it and is intelligent enough to prove that he wants it by organizing to get it."[56] Donald Van Vleet predicted "thousands of farmers" would soon be filing claims for refunds and joining the Association.[57]

What was the reaction of the Department of Agriculture to the efforts of the Association? At first the Department of Agriculture was only dimly aware of their activities. But as the Department officials heard more about them, they attempted to counter the activities of the Association in various ways.

One USDA tactic was to attempt to co-opt its critics, to persuade them to become allies. Officials of the Department made a strong effort to attract the Farmers Union leadership. Emil Loriks, Farm Holiday leader and South Dakota Farmers Union president during the 1930s, was contacted by Rexford Tugwell, Undersecretary of the Department of Agriculture, when they were both present at a meeting in South Dakota, and introduced to President Roosevelt. Loriks had the opportunity to spend an hour with the President following their meeting and was greatly impressed. When the Recovery Association asked Loriks if he would head up their program in South Dakota, Loriks was not interested. Loriks later became a regional administrator for the Farm Security Administration.[58]

Another tactic was to threaten their critics. Paul Appleby, Assistant to Secretary Wallace, wrote to an Iowa Recovery Association recruiter charging that he suspected him of deceiving farmers by claiming to be able to obtain a refund for them.[59]

In January of 1937 the USDA alerted the Postmaster General to a complaint by the state's attorney of South Dakota against a representative of the Farmers Process Tax Recovery Association. According to the complaint, the Recovery Association was making false and misleading statements: that the Department of Agriculture said "tax was collected by reducing the price of one hundred pounds of live weight by the amount of tax," and the "consumer, according to compilation of figures by the United States Agriculture Department, never paid any part of the tax."

The Department writers commented: "The whole scheme appears to be merely a device to mulct hog producers." The Department especially singled out Melvin Hoard of South Dakota, an independent tax consultant who had links with the Iowa Farmers Process Tax Recovery Association, and who had been charged by the South Dakota Attorney General, C.C. Dayton of southeastern Iowa.[60] Apparently, two investigations were made. At the end of the second investigation, Post Office officials sent a note to the Department of Agriculture saying that their investigation had found no evidence that would support a criminal proceeding.[61]

The USDA directed AAA committeemen and county agents to oppose the Association. When Lee Gentry, the chairman of the Illinois Agricultural Conservation Committee, asked how to deal with a committee chairman who was trying to enlist non-participants in the corn-hog program in the recovery campaign, the USDA replied that the offending committeeman was "subject to removal."[62]

There is evidence that county agents received instructions to oppose the activities of the Recovery Association. Some farmers wrote that county agents thwarted their efforts to obtain proof of the hogs that they sold during the years when the processing tax was in effect. John Maier of Elgin, North Dakota, wrote that when he talked to his county agent he would not give him the sales slips of the hogs he sold to packing companies. He had turned these slips in to the office, when he enlisted in the AAA corn-hog program. The county agent just gave him a copy and he would not sign it.[63] Andrew Hoganson of Hopper, Nebraska, wrote that he had contacted his county agent in order to get the figures for the year he was under the AAA program; the county agent told him "they got orders to keep them."[64]

A still more powerful tactic of opposition used by the Agriculture Department was a press release that was sent to rural newspapers across the nation. It was misleading in that it intimated that any efforts of groups such as the Farmers Process Tax Recovery Association were fraudulent. It warned farmers against groups soliciting funds from farmers to help them obtain refunds of the processing tax. On January 11, 1938, Martin G. White, Solicitor of the Department of Agriculture, issued a warning to farmers:

County AAA office. (Courtesy of Library of Congress, Washington, D.C.)

"In connection with this matter it should be emphatically stated that there is no provision of law which authorizes or allows any refund of processing taxes to a producer unless such producer was the actual processor and himself paid the processing taxes to the collector of internal revenue and did not pass such taxes on to the consumer.... Farmers should be warned not to allow themselves to be mulcted of funds upon such promises."[65]

The USDA sent out the statement, even though Representative William Lemke had introduced a bill in 1937 for refunding the processing tax to hog producers. But the Department of Agriculture continued to distribute the Mastin White statement for publication in an attempt to discourage popular support for recovery and the Lemke bill.[66] During the next few months the White statement was published in a number of newspapers throughout the country. Numerous articles based upon the statement were also published.[67]

The Department of Agriculture sent the White statement to the packing companies, telling them to warn farmers against efforts to recover the processing tax. Furthermore, they were told it was useless to provide farmers with evidence of their hog sales. One South Dakota farmer wrote that he got a letter from the manager of Swift and Company at Watertown, South Dakota, saying that "wires and letters that we have received from Washington indicate there is very little possibility of any bill being enacted."[68] A Minnesota farmer wrote that when he asked for his records, the treasurer said, "George, the Packers paid the tax, and not you, they would be the ones to get the money if returned, but if the government intended to give it to the Farmer, it would be made known publicly, so if you don't have other sales to look for ... don't bother about ours."[69]

The Farmers Process Tax Recovery Association officials angrily countered the Department statements. Donald Van Vleet declared that the USDA propaganda was put out with a "deliberate intent to hurt our organization." He said that every publication to whom the officers had explained the situation had come out with the information that the organization was legitimate and had the right to organize and fight for the recovery of the money.[70]

Edward E. Kennedy was also angry about the White statement and the newspaper coverage it was given. He told White that since the processors had worked for a bill that permitted them to reclaim part of the processing tax, it was "only right that the farmers could also organize to get back the processing taxes which they bore." He said the Association was a legitimate group, with officers who were well-known and honorable men. One of the purposes of the Association was to petition Congress for the enactment of legislation for "refunds to hog producers who bore the burden of the Processing Tax." White readily agreed that NFPTRA efforts were "entirely proper and legitimate," and he said that he was not challenging the right of the farmers to "organize as they see fit ... and use the force and power of the organization to petition Congress and get the Processing Tax Refund legislation enacted by Congress."[71]

Kennedy's challenge made White retreat somewhat from his earlier statement, but it was still put out by the Department of Agriculture after this. Kennedy and Lemke, in addition to their interest in recovering the process tax, continued to work for cost of production. They often united the two causes. Writing at a later date, Kennedy stated that he had been using the demand for the return of the processing taxes to keep the Department of Agriculture from attempting to support new agricultural legislation by processing taxes.[72] To Van Vleet, Kennedy commented, "I have followed the strategy of raising the question of our demand for the refund of the last unconstitutional processing tax and demanding that the last bill be paid to the farmer before another bill is incurred. And [I am] offering a substitute that will get the farmer the cost of production and not cost the Federal Treasury any money."[73] The cost of production bill was very much on the minds of both Kennedy and Lemke in the closing months of 1937. Lemke wrote to a Minnesota farmer that he was doing all he could to get a "real agricultural bill passed.... We want cost of production substituted for the bill that is before us.... As a matter of fact, Mr. Kennedy and I helped write it [the cost of production bill]."[74]

Prior to introducing a new resolution on the processing tax, Kennedy wrote that he had been working with Lemke on the preparation of material to go with the resolution. He had also hired some additional help to assist him. They had searched the Library of Congress, and the Congressional Record and committee hearings dealing with the subject over several years.[75] In late January Representative Lemke presented a speech in the House of Representatives in support of the resolution authorizing the refund to the producer of the processing tax on hogs. He said that Congress acted promptly to refund the processing taxes to the processors and to the distributors but did not give refunds to the farmers, "who it appears really paid the processing taxes...." Lemke also said that he had the assistance of Edward E. Kennedy in the collection of the facts. "I am sure Members of Congress who are familiar with farm legislation will agree with me that Mr. Kennedy is one of the best authorities and one of the best informed men on farm legislation and the needs of farmers in this nation."[76]

Kennedy mailed Van Vleet 200 copies of Lemke's speech, and ordered 2,000 more.[77] Lemke also sent out a letter mentioning his activities in connection with legislation for the return of the processing tax.[78]

At the same time, Kennedy was sending copies of Lemke's speech to members of the Farmers Process Tax Recovery Association. He also told the leaders of the organization about plans for committee hearings. He said that he and Lemke had arranged with the chairman of the Subcommittee of the House Agricultural Committee for hearings on the Resolution. These hearings would be held in about two weeks. Kennedy suggested that the entire board of directors, as well as state managers, should be there for the committee hearings.[79]

When the leaders of the Farmers Process Tax Recovery Association gathered in Washington for the Senate and House hearings on their proposed legislation, they also held a meeting of the board of directors. Meeting in Kennedy's office the afternoon of March 19, they decided to sign a contract with Gurney. They also decided that the responsibility of filling the time allotted to them on radio station WNAX would be given to Kennedy.[80] Although the minutes of the meeting do not mention it, later letters by Van Vleet indicated that he was very much opposed to accepting Gurney into the organization because Gurney charged lower rates than the Farmers Process Tax Recovery Association did. But he was out-voted.[81]

Preparing the way for the Washington meeting, Kennedy's secretary, Mary Puncke, reported that Representative Lemke would speak first before the Subcommittee of the House Agriculture Committee. Then she continued, "The Department of Agriculture is working hard to stop, if they can, the hearings on our bill."[82]

The chairman of the Senate Committee on Agriculture and Forestry was Ellison D. Smith of South Carolina. There were several other members on the committee from the South.[83] The committee heard arguments from other farm groups seeking the return of the processing tax on cotton and tobacco, as well as the Farmers Process Tax Recovery Association's arguments for the return of the hog processing tax.

Many of the farm groups believed Congress had intended that the benefits of the processing tax go to the farmer. But the intent of Congress had not been carried out. Speaking at the Senate Hearing on the bill submitted by the Farmers Process Tax Recovery Association, Kennedy commented that he felt Congress had intended that tax be passed on to the consumer and that benefits payments made from that fund would be transfers from the consumer to the farmer who participated. But instead the tax was deducted from the market price paid to the farmer for his hogs.[84]

To support his arguments, Kennedy submitted a document prepared by the Bureau of Agriculture Economics titled "Analysis of the Effects of the Processing Tax Levied Under the Agricultural Adjustment Act." According to this analysis, retail prices of hog products in 1934 and 1935 were no higher than they would have been if the tax had not been in effect, processors' margins were widened by about the amount of the tax, and retailers' margins were not affected by the tax; therefore, "live-hog prices were lower by about the amount of the processing tax."[85] Kennedy also stated that hog packers had testified at a Senate and House Agricultural Committee Hearing in 1933 and 1935 that they were charging the amount of the tax back to the hog producers in a lower price for their hogs.[86]

Kennedy said, "The amount of this tax places a tremendously heavy burden upon the hog producers." It was 50 cents a hundredweight in November of 1933, then rose to $1.00 per hundredweight, in December of 1933. In February of 1934 it was increased to $1.50 per hundredweight and in March of 1934 it reached $2.25 per hundredweight and stayed at that point until the processing tax was declared unconstitutional in January of 1936.[87]

He also referred to the passage by the Senate of bills to refund penalty taxes to producers under the Bankhead Cotton Control Act, the Kerr-Smith Tobacco Act and the Potato Act. He continued, "I recall in the debates on this floor of the Senate at that time when the question was raised, 'does this take any new money out of the Treasury?' and the answer was made on the Senate floor that this did not take any new money out of the Treasury. It only provided for refunding money that had been illegally extracted for these producers and put in the Treasury.... This same thing is true under this bill."[88]

Senator Frazier of North Dakota, who had introduced the Hog Processing Tax Refund bill into the Senate, asked Kennedy if there was any claim made that the processing taxes on wheat and wheat products had been paid by the farmer also. Kennedy said the wheat tax had been passed on to the consumer.

Senator William Bulow, a Democrat from South Dakota, asked if there was any present way a farmer could file for return of the processing tax. Kennedy answered that, according to the earlier act, the farmer must actually have paid the tax to the government and gotten a receipt for it in order to file for a refund. The hog farmer did not do this unless he was a processor also.

Senator Gillette commented that he felt as a matter of equity the nonparticipants in the old AAA were entitled to "reimbursement for any injustice that they suffered by reason of the imposition of the tax." But he had serious questions about the bill because he was not certain that farmers would be able to prove they had suffered a loss.[89] Kennedy replied that the law which permitted persons to apply for refunds under Title VII of the Revenue Act of 1936 set forth the requirements for evidence and presumption for proving claims. He suggested that farmers should follow the same rules.[90]

Next, Representative Lemke spoke before the Senate Committee. He said he had introduced a similar bill before the House and all the bill did was give the farmer the same opportunity as Congress gave to the packers and processors.

Senator Frazier asked, "Has any estimate been made as to the amount of money that would be refunded or subject to refund if this bill passed?" Lemke replied, "That is largely a question of speculation," but concluded that the total amount, if it were all recovered, was $361,000,000.[91]

The next person to testify was Donald Van Vleet, who quoted Geoffrey Shepherd of Ames, an economist who had prepared a series of articles on the hog processing tax, again saying the farmer bore the burden of the tax.[92]

Van Vleet was followed by officers and presidents of the Farmers Union, coming from Iowa and nearby states. A.J. Johnson said farmers were aware that Congress had made provisions for processors so that if they could show they bore the burden of the tax, they were able to submit their claims. John Erp said farmers were of the opinion that the tax was taken from them, "because of the low hog prices that we had during that period." Harry Parmenter of Nebraska talked about the low prices farmers had been receiving for their products; low prices which had driven many farmers into bankruptcy. Congress, he said, having given relief to the processors, certainly will recognize that it is only fair to grant "equal restitution to the substantial citizens of the country, the basic industry of this great nation, the one industry that everyone in the United States had to depend upon to live." Parmenter further thanked the senators for the opportunity to appear there on behalf of the "drought-ridden, grasshopper-ridden, insect-ridden, mortgage-ridden farmers in the Midwest."[93]

Emil Loriks, president of the South Dakota Farmers' Union, also spoke before the committee. He said that he favored the return of the hog processing tax to the farmers, but he realized that it would be a difficult task to get these tax refunds back to the producers. He felt that the government should provide the vehicle, the machinery necessary for returning these taxes to the producers. He did not believe that the farmers should be put to added expense to secure these refunds.[94]

The Senate Hearings were followed by shorter hearings before a special subcommittee of the House Committee on Agriculture. The House subcommittee heard testimony relating to tobacco, cotton, potato and hog processing taxes. A representative of the Department of Justice stated that the amount involved in the cotton, tobacco and potato refund claims was around $6,000,000.[95] This was much less than the estimated cost of the processing tax refund on hogs.

Representative John Flannagan, a Democrat from Virginia, said, as he understood it, the Department of Agriculture was against the passage of this legislation. A Justice Department representative who had been asked to testify said, "Not against the legislation as it originally was proposed." Actually, he stated, his reason for coming to the Hearings was to see that a bill was passed to prevent the filing of a great volume of claims against the government. He believed that the government was likely to lose suits brought by cotton or tobacco growers, and so the Justice Department was recommending the passage of legislation for refunds to these groups. "It is advised that the Government will have to pay the losses if suits are brought in court anyway."[96] The Internal Revenue Bureau would prepare a form and send it to all litigants for filing, "the thought being that most of them will be paid off, and therefore, in those cases, litigation will not be necessary."[97]

Following these presentations, Kennedy introduced the arguments of the Farmers Process Tax Recovery Association. He said the Treasury Department in 1937 had issued a statement that in order to avoid multiple suits for refunds of cotton, tobacco, potatoes and other commodities, remedial legislation should be provided for the orderly refund of these taxes. He hoped to include the hog processing tax along with them.[98]

Johnson, Erp and Van Vleet spoke briefly at the subcommittee hearings. John Vesecky of Salina, Kansas, President of the National Farmers Union, also spoke before the committee. Declaring the AAA had worked fine for wheat farmers but not for corn-hog producers, he strongly supported the resolution. Both the "fellow who participated and the fellow who did not participate should be given the same privilege to show that the tax was taken from the price of hogs." When asked if the reason for the lowered prices was because the packers did not want to cooperate, he replied, "They said they did not want to go into the program.... This was the first opportunity the farmers had, and the packers got sore about it and threw their opposition to the farmers getting a break. They said they did not want anything regulating their business.... And they did everything they could to see that it was not a success."[99]

Kennedy and Johnson were questioned as to how much money had been collected from farmers who filed their claims for refunds of the hog processing taxes through the Association. They said about $5,500 had been collected.[100]

Congressman Lemke was the last to testify before the 1938 House Subcommittee. He said he had offered the amendment after "hundreds of farmers" had asked to be given an equal chance with the processors, who did not pay the tax.[101] Lemke said that 17 out of a hundred received benefit payments, and 85 percent did not. (Lemke's mathematics seem a little faulty here.) The 17 cut down their production, and therefore he believed both those who participated in the AAA program and those who did not should be permitted to apply for the tax refunds. He also suggested that the committee report the hog processing tax and the cotton tax bills together as one bill.[102]

Following these hearings, Chairman Smith inserted a letter from Secretary Wallace dated September 29, 1937. Wallace opposed the refund of the processing taxes because the hog producers received higher prices following the corn-hog adjustment program. He stated: "It should also be kept in mind that many of those producers who remained out of the corn-hog adjustment programs took advantage of the situation by increasing their production while cooperating producers were curtailing and, as a group, the non-cooperating producers were able to sell many more hogs at a price then prevailing then they would have."[103]

On March 17, 1938, Wallace informed the House Agriculture subcommittee that in order to avoid appeals to the Supreme Court and because the administration was beginning important new programs relating to the regulation of the marketing of cotton and tobacco under Title III of the Agricultural Adjustment Act of 1938, "this Department will not oppose the enactment of legislation which will authorize the making of refunds of the amounts collected as taxes under the Kerr Tobacco Act, the Bankhead Cotton Act, and the Potato Act."[104] Thus, the Department of Agriculture permitted the granting of refunds to cotton, tobacco and potato producers but continued to fight the refunding of processing taxes to hog farmers.

On March 25 the House Agriculture Subcommittee approved the hog processing tax refund bill with the amendment (which the House subcommittee had suggested) to include the refunds of the processing tax to farmers who participated in the AAA program.[105] Representative Mitchell and Gilchrist voted for the bill, and Flannagan voted against it.[106]

Upon returning home, the officers of the Farmers Process Tax Recovery Association sent a letter to all the state managers and county committee men expressing satisfaction with their successful hearings before the congressional agriculture committees. "All witnesses who testified in favor of the bill did so in a united way.... Some very favorable sentiment has arisen recently in favor of the bill from the states where the sign-up was large."[107]

Members of the Farmers Process Tax Recovery Association were enthusiastic and optimistic, perhaps too optimistic. One Iowa farmer wrote: "Enclosed I am sending one dollar for my membership fee for the year. I have watch[ed] the paper every day for the out come of this meeting in Washington but so far have not seen a thing. If you have any literure [sic] on the meeting I would sure appreciate hearing from you.[108]

Some of their friends and neighbors adopted a "wait and see" attitude. For example, another farmer wrote to the Association headquarters: "Enclosed find a check to pay my dues for the 1938 Process Tax Recovery Assn. There is a lot of farmers that would like to have the tax returned to them, but they have been fooled so many times they are afraid, and they have been told by so many that it is no use. I have tried to talk to these fellows and they are interested but are afraid they will lose what they put in it and say they haven't got the money to lose. I am glad that there is still a few that have *guts* enough to fight for what is there's [sic]."

But that winter recruitment slowed down. Van Vleet wrote to Kennedy that, "Business has not been very good the last ten days."[109] Kennedy responded by encouraging the leaders to new efforts. "Write to each of your two Senators and ask each of them for a copy of the hearings. At the same time ask them to help get this Bill enacted into law as soon as possible."[110]

Some opposition to refunding the hog processing tax began to appear in Congress. In April Representative Jerry O'Connell of Montana made a speech and then inserted an extension of his remarks in the *Congressional Record*. He said that the organizations who worked for the refund of the hog processing taxes were rackets and they were run by the "services of a radio station, legal talent, and a former farm official who was repudiated by the membership of his organization in the election of officers." He said they were "selling legislation," and he would do everything possible to keep them from profiting at the expense of the farmers. Furthermore, if a processing tax refund bill were passed, since very little had been returned from the packers windfall taxes, he would recommend that the money to pay the hog producers come from the packers and not the Public Treasury.[111]

By June Lemke was in North Dakota campaigning in the primaries for nomination and re-election. Kennedy in Washington helped by securing a number of endorsements from Senators and Representatives and writing newspaper releases.[112]

He also met with D.B. Gurney, who came to Washington to check on the progress of the refund bill. Gurney wrote a number of letters to contributors to his organization, saying that he had been to Washington and learned "at first hand the interest and knowledge of the pending

legislation.... I am naturally pleased at the progress that has been made....
We must consider, of course, that federal legislation moves slowly, regardless of the merits, but ... it is all very gratifying."[113]

On June 25, 1938, Congress passed appropriations providing for the
refund of cotton, tobacco and potato processing taxes that had been levied
under the Bankhead, Kerr and Potato Acts. But no action was taken on a
bill appropriating funds for the repayment of the hog processing taxes.[114]

About this time, Van Vleet resigned, citing differences between him-
self and Kennedy over the direction of the Association. A major issue had
been the alliance with Gurney. Work in the Farmers Process Tax Recovery
Association continued under the leadership of John Erp, president, and A.J.
Johnson, secretary. Johnson carried on much of the work of the office in
Des Moines, with the help of Farmers Union secretary Helen Holehan.[115]

This winter, membership in the Farmers Process Tax Recovery Association was increasing. In January of 1939 A.J. Johnson was "so completely
snowed under with claims and correspondence that it was impossible to
get around to" writing replies to all his correspondents.[116] The group continued to experience periodic highs and lows of enthusiasm among its
supporters.

Another development that affected the Recovery Association was the
formation of another farm organization. On January 16 Robert Spencer,
Chairman of the Committee on National Organization, announced a meeting to be held in Peoria, Illinois, to "set up a new farm organization." The
announcement stated that E. E. Kennedy would be there both days of the
proposed meeting, and that the purpose of such an organization would be
to "help Kennedy and his co-workers in Washington get the Cost of Production bill, the Frazier Lemke Refinancing Bill and the Hog Processing Tax
Recovery bill made into law this session of Congress."[117] It was hoped this
new organization (The Farmers Guild) would strengthen the efforts of
Kennedy in the campaign to recover the tax; however, it probably weakened the impact of the Recovery Association, as many of the strongest supporters of the tax recovery plan left the Farmers Union to join the new
group. Thus they would have less impact on the Farmers Union, the only
national farm group where they had held some influence before.

On January 30, 1939, Congressman Lemke introduced a joint resolution, House Resolution 138, providing for the refund of the processing tax
on hogs. The bill was referred to the House Committee on Agriculture.[118]

Lemke expressed his anger at farm leaders who opposed his farm legislation. "While I have been fighting for that which I am confident is for
the best interest of the farmer in our state, and have been constantly for
the last three years, there are clicks with a stiletto in my back."[119] In another

letter he wrote, "Their [the farmers'] representatives have not stood solidly behind any real legislation. Some of them have rather been the chore boys of the Secretary of Agriculture.... They have represented the Department of Agriculture and not the farmers who they are supposed to represent. They have been looking for jobs or other favors from the Department rather than demanding justice. This is not true of all, but some of them."[120]

In March Lemke made a speech in the House concerning his cost of production bill and the processing tax. He stated that Secretary Wallace had forgotten to tell the members of Congress "the processing tax was charged back to the farmer in lower prices." The farmer paid the processing tax. "He forgot to tell us that his own Department had made a report that the farmer paid the tax in lower prices." The Bureau reported that the farmer who signed the hog and corn contract received $2.34 less per hundred pounds when the tax went into effect. Therefore, Lemke believed, the farmer paid $2.34 in processing tax in order to get $2.20.[121]

Hopes for passage of the cost of production bill and the hog processing tax refund bill dimmed when, despite their strong efforts, Lemke, Kennedy and others were unable to get the cost of production bill passed. They received a serious setback in their plans when the new Farmers Union leaders informed House Agriculture Committee Chairman Jones that they opposed the cost of production bill. To many observers it came as a surprise.[122] To Lemke and Kennedy, who had been complaining that farm organization leaders were becoming too friendly with the Roosevelt Administration, it probably came as a further confirmation of their own fears.

Postponing for a time any further action on the cost of production bill, Lemke and Kennedy took up the refunding bill. Hearings were held before the Senate Subcommittee on Agriculture in May of 1939. The members of the Senate subcommittee were Lynn Frazier of North Dakota, Chairman; and Senators W.J. Bulow of South Dakota, and Guy M. Gillette of Iowa. The hearing began with testimony by Kennedy, who asked that the committee also consider evidence presented the previous year. Secretary Wallace and others had said that it would be impossible to determine how much was owed the farmers. In reply, Kennedy submitted claims that farmers had filed with the Farmers Process Tax Recovery Association. These claims listed who sold the hogs, to whom they were sold, at what time and at what weight. Since Congress knew the prices of processing taxes that were charged at various times, it would not be difficult to determine how much any particular farmer had paid in processing taxes.[123]

Representative Lemke also testified on the bill. He said that he was in favor of the amendment, which he had submitted in the last term, providing for paying both the farmers who refused to sign up with the AAA and those who signed up with the AAA program.[124]

In his testimony, A.J. Johnson told the committee about the Farmers Process Tax Recovery Association. He said they represented 3,100 claims for a total amount of $1,344,690. This was on the average $433 per claim. Only four claims were for more than $5,000.[125]

Other testimony was given by Christian Grell, a member of the governing board of the Recovery Association; and Robert Spencer, the Recovery Association state director for Indiana. Spencer emphasized the small size of most claims. The average claim in the state of Indiana was $373. "Now what is a farmer going to do with a small claim like that? He cannot hire a lawyer. He cannot come down here for a private bill. He cannot pay fees. He cannot furnish the kind of proof a businessman with a staff of clerks and bookkeepers can. But his claim for his little $375 tax refund is as legal and means as much to him as a $375,000 claim means to some company with a thousand stockholders, and he has as much right to it as any corporation that has paid taxes under an unconstitutional law."[126]

The hog processing tax refund bill was reported favorably out of the Senate subcommittee in 1939. Also, a bill was passed, which the Farmers Process Tax Recovery Association had favored, extending the time whereby the few farmers who had also processed their hogs could file for refunds. The earlier law for requesting refunds had expired in June of 1937 but it was extended to February 1, 1940.[127]

But the processing tax refund bill did not get out of the House Agriculture Committee. In late July Lemke wrote to a farmer in Iowa that the "House will not act because you who are interested in this have not been active enough at home."[128]

The following year, supporters of the bill again met discouragement. In late April a vote was taken in the House Agriculture Committee on the processing tax. On the first vote, the bill was approved by the committee, but the committee chairman called a recess so that several members who were not present could be notified. On the second vote, the bill was defeated by one vote.[129] Lemke reported: "Two Republicans, Hope of Kansas and Kinzer of Pennsylvania, voted with the Democrats to defeat us; four Democrats, Pierce of Oregon, Hook of Michigan, Polk of Ohio, and Pace of Georgia, voted with eight Republicans, making the final vote twelve for and thirteen against."[130]

(It is surprising to note that one of the Republicans voting against the hog processing tax refund bill was Representative Hope from Kansas, who earlier had spoken in Congress about the need for the packers to refund the processing tax to the farmers.)

The team of Lemke and Kennedy decided to try another tactic to achieve the passage of the hog processing tax refund. Lemke brought up a Senate Joint Resolution on the processing tax; this was referred to the House Agricultural Committee, and an amendment was added to it by Congressman Pace of Georgia who wanted a cotton tax refund bill also passed. Now with southern support, the bill had enough votes in the committee to pass — 15 out of 25. But the chairman of the House Agriculture Committee, Marvin Jones, ruled that the motion to report out the bill was out of order because a similar bill had already been considered in the same session of Congress. Many of the members thought that Jones' ruling was wrong but hesitated to vote against him because that would seem like a personal affront. So again the processing tax refund bill was defeated. (Kennedy believed that Jones' adverse rulings on the processing tax refund bills were because he had been promised a federal judgeship if he followed the Administration's wishes.)[131]

In May Lemke was scheduled to begin his campaign for re-nomination to Congress in North Dakota. He wrote to one of his supporters that because of several bills, including the hog processing bill, which would be coming up within the week in the Agricultural Committee, it would be impossible to make the speaking engagements scheduled for the first week of the campaign.[132]

Nothing was accomplished, however, and Lemke journeyed on to North Dakota for the campaign. His secretary in Washington sent notes to those who asked about the hog processing tax refund bill stating that things were at a "standstill for this session of Congress. There has been a great deal of activity this session in behalf of the Act, but Washington is very much taken up with war hysteria, and the chances for passage this session seem to have passed completely out of the picture."[133]

In the campaign for re-nomination to the House of Representatives, Lemke won easily, but his friend, Senator Frazier, was defeated by former Governor Langer. Lemke was then urged to campaign for the Senate against Langer in the fall elections. He did so, but this proved to be a mistake because Langer defeated Lemke in a closely fought battle.[134]

Lemke had fought long and hard for the passage of the hog processing tax refund, but now he retired to private life in North Dakota and the fight was carried on by other leaders.

In the fall of 1940 the Farmers Process Tax Recovery Association tried a new method; they sent a proposal to the Senate Appropriations Committee calling for the refund of the processing taxes. The bill was considered by the Committee, and Kennedy presented some material for the group to study.[135] In October an amendment introduced by South Dakota's

Senator Gerald Nye failed by one vote to become part of an appropriation bill.[136]

But the Recovery Association leaders did not give up. In February of 1941 Representative Fred Gilchrist of Iowa and Senator Chandler Gurney of South Dakota introduced bills calling for the refunding of the hog process tax.[137] They were determined to keep on trying. A.J. Johnson wrote one of the members of the Recovery Association, who had asked about the possibilities of success of their endeavor: "I know nothing that would lead me to believe that this tax will not be refunded in time. The officers of our Association realized at the beginning of this fight that it would be a lengthy one."[138]

There were problems. Other issues were attracting more attention. Congress was again considering measures related to the start of World War II. Kennedy wrote, "Gilchrist feels very sure that we can get hearings just as soon as the lease lend [sic] bill is out of the way." And money continued to be scarce. In March Kennedy asked, "Are we going to have enough to continue the fight through this congress?"[139]

In May there were more hearings on the hog processing tax refund bill before a subcommittee of the Senate Committee on Agriculture and Forestry. Guy Gillette of Iowa was the chairman, and other members were Berkeley Bunker, a Democrat from Nevada, and Raymond Willis, an Indiana Republican. In a letter to the subcommittee, Paul Appleby, Acting Secretary of Agriculture, declared that the money from the processing taxes had all been spent, and to pay the claims for hog producers would place the burden of such refunds on the public.[140]

Speaking before the Senate Committee, Congressman Gilchrist said that the Treasury had about $22 to $24 million still "in their pockets" from the hog processing tax. And the farmers paid that tax. "In spite of what my good friend Mr. Appleby says, I call the attention of the committee to this brochure [written by A.G. Black of the Bureau of Economics in the Department of Agriculture in 1937] which will absolutely refute the statement so made."[141]

The subcommittee also heard from Kennedy, Case and Gilchrist. Kennedy stated that the purpose of the BAE booklet was to defend the treasury against the processors.[142] South Dakota Representative Francis Case read a statement detailing the many instances in which the Federal government had refunded illegal taxes. It was titled, "Uncle Sam Has Refunded Illegal Taxes for Just About Everybody Except."[143]

Representative Gilchrist dismissed the Department of Agriculture claim that the farmer who did not cooperate with the AAA made money while others cut their production by saying that the prices for hogs were so low that no one made money. Gilchrist felt that the money had been stolen from the farmers, and one of the Ten Commandments is "Thou shall not steal."[144]

Senator Chandler Gurney criticized Acting Secretary Appleby's statement page by page. He also stated that the farmers in his area knew definitely that they were entitled to a refund. There had been literally thousands of meetings in Nebraska, Iowa, Minnesota and South Dakota concerning the processing tax. (Although Senator Gurney did not state this in the hearing, his father, D.B. Gurney, had organized and led many of those meetings.)[145]

On the last day of the subcommittee hearings, a representative of the Commissioner of Internal Revenue testified at the request of Senator Willis. Senator Gillette asked if there had been any claims on the processing tax which had been rejected. The representative said about $98,000,000 in claims had been rejected. These claims were filed by processors. Claims filed by producers had not been tabulated, since they were not considered proper claims.

In June of 1941 the U.S. Senate unanimously passed the Hog Processing Tax Refund Bill. The House Committee on Agriculture assured the backers of the bill that they would report the measure to the floor of the House within a few weeks.

Exultantly, Kennedy wrote in his *Washington Letter* of June 25, 1941, that "farmers should be getting evidence of sale together" and could consult the Farmers Process Tax Recovery Association for expert help in filling out their claims for submission as soon as the legislation was passed.[146]

The members of the Farmers Process Tax Recovery Association again felt that they were on the very edge of victory. But a lengthy letter submitted to the Chairman of the Committee on Agriculture by Herbert E. Gaston, Acting Secretary of the Treasury, was to destroy these hopes. Gaston wrote that costs of the hog processing tax refund bill would be much greater than the proponents of the bill had claimed, that all of the hog farmers would claim their refund, not all claims filed by the processors had yet been adjudicated, and it would "create a liability on the Government to 'refund' to hog producers that approximately $100,000,000 of hog processing tax imposed but never collected."[147]

That fall, things began to wind down. Kennedy sent a proposed reply to Gaston's letter to Congressman Gilchrist in September. He also stated that he would have to be absent from the city for a short period.[148] At this time Kennedy was involved in helping farmers in Ohio protest a Department of Agriculture Wheat Penalty Tax.[149] In October Congressman Gilchrist sent a letter to Kennedy from his hospital bed in Rochester,

Minnesota. He wrote, "I am disturbed about the hog processing tax bill because I cannot be there to give it the attention that I want to give it."[150]

In January of 1942 a member of the Recovery Association wrote that he had received word from Congressman Gilchrist that the chairman of the subcommittee had not filed his formal report with the full committee.[151]

At this time Christian Grell, the new director of the Iowa division of the Recovery Association, came up with a new twist to their appeals for recovery of the hog process tax. He suggested that the organization ask for repayment of the tax in defense bonds. He thought the new suggestion might put pressure on the House Agricultural Committee so that they would release the processing tax resolution to Congress and it could be voted on. It might also be a means of encouraging new membership in the Association.[152]

Grell sent copies of his resolution to Kennedy[153] and Johnson. A member of the Recovery Association discussed the proposal with Representative Gilchrist. Gilchrist had written to some of the Association members that "the strenuous times" had forced the House Agricultural Committee to work on other matters. But when the new idea was presented, Gilchrist said that the committee and the public might agree to return the tax to the farmers. "In fact he seemed more optimistic than I have seen him for some time."[154]

The new plan, however, did not save the proposals to recover the hog processing tax. The bill died in committee.

In February of 1942 Congressman Robert B. Chipperfield wrote to one of his constituents who had asked concerning the tax refund bill that although he, personally, favored the bill, he feared it was opposed by the Administration.[155]

About this time the leadership of the Farmers Process Tax Recovery Association gave up. One by one the group had lost its leaders—Donald Van Vleet, D.B. Gurney, William Lemke, Edward E. Kennedy—until finally the only one left was A.J. Johnson. Johnson kept the records of the group for years, perhaps hoping that the group might be revived again. But it never was. It had been a long struggle, and the hog farmers of the Middle West and their representatives had put up a valiant fight, but they had lost.

6

D.B. Gurney and the Radio Campaign

In 1938 in South Dakota another leader, D.B. Gurney of Yankton, spoke out against the New Deal and encouraged farmers to enlist with him in an attempt to recover the processing tax. He recruited farmers through his noontime radio broadcasts over radio station WNAX.

Many farmers in South Dakota were angry because their goods were selling for such low prices. Although the prices rose a little because of New Deal programs during the 1930s, they still were not as high as they had been before the Depression.[1]

Many South Dakota farmers who had borrowed to expand their production found that the low prices they were receiving for their crops were not enough to help them pay their debts.

The debt problems of the South Dakota farmers were reflected in the many farm foreclosures and bank failures. During the period from 1921 to 1932, 31,419 farm foreclosures had been instituted, involving nearly 19.6 percent of the farm acreage on the tax rolls.[2] Between 1920 and 1933 about 71 percent of all state banks failed.[3]

Extreme drought struck parts of South Dakota in the 1930s. Eastern and central South Dakota suffered from low amounts of rainfall — approximately 12 inches in 1933. Western South Dakota received seven inches of rainfall in 1931 and 13 inches in 1933.[4]

Accompanying the drought came plagues of grasshoppers. They consumed the crops that the drought weakened, attacking various areas of the state throughout much of the thirties.

Besieged by debts, drought, Depression and grasshoppers, South Dakota farmers were willing to listen in January of 1938 when D.B.(Deloss Butler) Gurney of Yankton, South Dakota, president of Gurney's Seed

Farmers and small towns suffered the effects of the Depression. (Dorothea Lange photo, Courtesy of Franklin D. Roosevelt Library, Hyde Park, N.Y.)

Company and WNAX radio station, initiated a campaign that promised to be of financial benefit to the hog farmers of the area. Gurney was known as a prominent state Republican and a concerned public citizen who in the early 1930s had called for and received hundreds of railroad carloads full of local aid for drought victims in Arkansas, South Dakota and Nebraska.[5] He had also been a leader in calling for a special session of the state legislature to pass legislation to provide loans for farmers and ranchers to purchase feed for cattle because of the shortage of feed from the drought of the preceding year.[6]

The campaign which Gurney outlined in January of 1938 during his noontime farmers' hour over WNAX was a program for the recovery of the hog processing taxes which had been levied from 1933 to 1935 under the Agricultural Adjustment Act.

Gurney told the farmers that the prices they received for their hogs had been lowered by the packers by the amount of the processing tax. Thus, they had paid the tax.

Gurney knew that an Iowa group had formed and called themselves the National Farmers Process Tax Recovery Association. They had first

Drought-stricken landscape in South Dakota. An Arthur Rothstein photo. (Courtesy of South Dakota State Historical Society Archives, Pierre, S.D.)

tried appeals through the Commissioner of Internal Revenue. Then, when that failed, they decided to try to get a bill passed through Congress providing for their reimbursement.[7]

At the request of the Recovery Association, Representative William Lemke and Senator Lynn Frazier of North Dakota introduced legislation calling for return of the hog processing taxes to the farmers. The bills were introduced in August of 1937. On January 24, 1938, Congressman Lemke spoke in the House of Representatives in support of Senate Joint Resolution #202 for refunding the hog processing tax.[8] Gurney had long supported Lemke and his legislative program.[9] He knew about the National Farmers Process Tax Recovery Association and other groups that were organizing at this time and seeking legislation for return of the processing taxes that had been levied on cotton and tobacco.[10] So, the idea of forming an organization to promote the return to the farmers of the hog processing tax was not new when D.B. Gurney urged his radio listeners who were corn-hog farmers to gather up their hog receipts from November of 1933 through January 6, 1936, and enroll with him in a program to get their processing taxes returned. Gurney was a showman, an entrepreneur,"[11] and he used his enthusiasm and persuasive abilities in the struggle for the return of the processing tax.

Gurney said the farmers should enroll in his program, send in their receipts, pay a small fee, and together they would try to get their money back.

There wasn't much that Gurney could do about the drought or the grasshoppers, but he may have felt that through his campaign he could help the farmers benefit financially. Certainly, if the farmers' lot improved, it would also benefit him and his seed company. Even if the processing tax were not returned, the effort to obtain its return would be a good public relations move for the seed company.

So Gurney began to suggest to his listeners on his noon hour farmers' program that they collect their hog sales receipts from the years when the AAA corn-hog program was in effect and send them in to him at Yankton, South Dakota, and enroll in a program with him to attempt to get their money back. He really did not have much organization. The letters came either to radio station WNAX or to Gurney's Seed Company, and then were passed on to him. Gurney used the services of a few secretaries at Gurney's Seed Company to open the mail and record the payments, generally answering people's letters himself.

Gurney was regarded as a friend by many of his listeners. Matt Fellers of Bluffton, Minnesota, wrote, "Listened to WNAX station every day. Please send me all papers for corn hog collection and all necessary information how to get the money."[12]

R.H. Cook of RD 7 Menomonie, Wisconsin, commented, "We listen to you talk over the Radio at 15 to one every day. Hope the refund will go thru, that would be a real God's blessing."[13] Charles Matz explained, "I have heard you so many times on the radio that I feel that I know you real well."[14]

Many other farmers wrote similar remarks in the letters they sent to Gurney. The letters came from listeners in South Dakota, Minnesota, Iowa, Wisconsin, North Dakota, Montana, Nebraska, Kansas and Missouri.[15]

At that time radio station WNAX had one of the highest towers in the Middle West, and its broadcasts covered a five state range.[16]

South Dakota farm. (Courtesy of South Dakota State Historical Society Archives, Pierre, S.D.)

Some of the letters came from people who did not even have a radio; they had heard about Gurney's plan to recover the hog processing taxes from their friends or neighbors and wrote in to Gurney to ask for more information.[17]

Occasionally they would ask for sign-up sheets for the processing tax recovery program at the same time they were ordering seeds or baby chicks.[18]

One letter writer said he was sending his hog slips to Gurney and did not know if it would do any good, but he hoped it would because he needed a "lot of alfalfa seed and clover seed and this money would come handy, and if you get this for me you can sure expect a big order."[19]

Often the letter writers asked questions concerning Gurney's attempt to recover the processing tax. One of the most pressing concerns seemed to be regarding eligibility. Listeners wanted to know which farmers would get the processing taxes returned to them — those who signed the government corn-hog contracts (signers) or those who did not sign the contracts (non-signers)? The contract signers had received payments from the government for reducing their numbers of hogs, while the non-signers had not received any payments. At first the bill applied only to non-signers, and so Gurney informed his listeners and letter writers that only those who had not signed up with the AAA corn-hog program were eligible.[20]

A few farmers were quite indignant that the proposed bill would only include the non-signers. Joseph Shirer of Rockham, South Dakota, questioned: "There are certain things which I don't quite understand. What I would like to know is, why the signers are not entitled to there [sic] refund of the tax just as well as the non signers. I should think if a law is illegal for one it would also be illegal for another. The signer payed [sic] the tax just as well as the non signer."[21]

Gurney wrote to answer these concerns: "Personally, I believe that the signer and the non-signer are equally entitled to this, but the present bills before Congress do not provide for the signer. I have promised you over the air that after this campaign is finished that I will see that another bill is introduced, taking care of the signers."[22]

In a later letter Gurney explained to another correspondent, "I have asked to have the Bill amended, and Congressman Lemke (one of the introducers of the bill) has agreed to do it if I insist, but he says it might make a considerable delay, possibly defeat the present Bill if we added that amendment."[23]

To another farmer Gurney wrote: "... I want every hog producer to get everything that is coming to him and do not want to overlook anything at all where the farmer and hog producer is concerned."[24]

In March of 1938 the subcommittee of the Committee on Agriculture suggested that the signer and the non-signer be given the same rights for the refund of the processing tax in the proposed bill, and recommended that the bill be passed as changed. Within a day or two, Gurney was announcing the new policy over his radio program and in his letters.[25]

On March 28 Gurney wrote to Henry Foreston in Foreston, Minnesota: "I know this will be interesting to you and it is very pleasing to me because I have made a stand for this and have worked hard to put the signers on the same basis as the non-signers."[26]

Many farmers responded to Gurney's proposal by busily hunting up their records of hog sales. Some farmers had very good records. They quickly filled out and returned a standard form that Gurney sent to them indicating whether or not they had participated in the government corn/hog program from 1933 to 1936. They were supposed to list the number of hogs sold, the purchaser, address of purchaser, total weight and total tax paid on the hogs. The sheet informed the farmers as to what the processing tax was during each year. (It was $.50 per hundredweight in 1933, $1.50 per hundredweight for part of 1934, and $2.25 per hundredweight from March 1, 1934, to January 6, 1936.) Then the farmers were supposed to total up their claim and pay Gurney about 2 percent of the amount claimed as a filing fee.[27]

They also agreed that if the processing tax was refunded, Gurney would deduct 6 percent of the amount collected and forward the balance to the farmer.[28]

Many farmers filled these forms out from records they had kept or were able to secure from small packing houses.

Others had more problems. Some farmers said that they had recently moved or cleaned house and destroyed or misplaced their records. Would they be able to get duplicates?[29]

Warren Bros of River Falls, Wis., wrote that his wife had "burned up a lot of slips last fall, she said they were no good."[30]

Michael Traxinger of Herreid, South Dakota, expressed his concern that it was difficult to obtain new receipts, as a lot of the buyers were "not around anymore."[31]

Farmers began to try to get copies of their hog receipts from the packing houses and commission firms. Frank Kinney of Gary, South Dakota, had gone to see his packer at Watertown and asked for the slips, and "they turned me down cold."[32]

Max Fiegen of Dell Rapids, South Dakota, went to Morrell's, but they said it took too much time to look for the hog receipts, "so we didn't get them."[33]

E.A. Hoegh of Hampton, Nebraska, said he had contacted two firms, the Farmers Union Commission Company, and the Triangle Commission Company for duplicate slips of his hog sales. The Triangle Commission Co. wrote that the "government had given them permission to destroy their past records so they could not furnish ... a duplicate sales slip. The Farmers Union Commission Co. would not send them ... stating it was 'early enough to get them when the bill was passed.'"[34]

Ole Trooien of Hendricks, Minnesota, writing to Gurney, enclosed a letter from John Morrell and Company:

> Dear Sir:
>
> In answer to your letter of February 3rd. We noticed in a late issue of *Wallace's Farmer* a statement from the Solicitor of the Department of Agriculture that this activity with reference to refunding Processing Taxes is not legitimate so far as any Government Agency is concerned. Therefore, until we know that the Government is behind this we cannot go to the expense of looking up all the information that is being requested by a number of our shippers.[35]

R.A. Rae of Worthing, South Dakota, complained to Gurney that he went to the Morrell Packing Plant in Sioux Falls to get dates and weights, but they would not look up the dates and weights for him. He said they claimed they did not know anything about the processing tax, and that there were many people coming in asking for information on their hog sales during the years of the processing tax. "They had that article published in the *Wallace Farmer* to read to everyone to beware of collectors of the processing tax."[36]

Gurney explained that apparently a large number of people went into the offices and "it sort of disturbed their work and they were more or less peeved about it." However, he continued, "our attorney has been to these plants now and explained the situation and told them that we would handle it on a basis that would not cost them anything ... [and] would not disturb them in their work."[37]

Gurney then began to send "Form B" to those farmers who did not have all their sales slips and needed help in establishing their claims to refunds of the processing taxes. These forms had blanks at the top for the date, name and address of any person or firm to whom the farmers had sold hogs during the years of the processing tax. The farmer who filled out the form requested that the firms whom he had named would furnish duplicate sales slips or other evidence of hog sales to D.B. Gurney or his representative. Once returned, the forms were kept in the farmer's file for use when the processing tax refund was finally passed.[38]

As Gurney assured his audience and letter writers, "We will have a very large number [of claims] in each of the commission houses and the packing companies in the various cities, and we will send our own help there and do the checking and not disturb them in their work. Or, we will pay their own employees to get this for us."[39]

Apparently, this still continued to be a matter of concern, because in March of 1938 Gurney wrote Paul Kantor of Luzerne, Iowa: "I have asked over the air that the grower and producer of the hogs does not bother the people to whom they sold their pigs at the present time...." Gurney expressed his belief that the packers and commission houses should be in favor of the passage of this bill so that the farmers can get the money back: "It would make better customers of the farmers for the people who expect to make their living from the stuff that you grow."[40]

While the non-signer had the problem of trying to find his hog receipts, the signer had a different problem. He had already found his corn-hog receipts and submitted them with his corn-hog contract to the government. Many of the farmers who went to their county agents to ask for their receipts or copies of their corn-hog contracts, which contained the same information, were turned down. Jno. Mueller of Clearwater, Nebraska, told Gurney, "I was to my Co. Agent and he stated that thus far he had no right to issue me my original sales slips." The County Agent also gave him a copy of the statement by Martin G. White, Solicitor, United States Department of Agriculture, which had also been referred to by some of the packers. In this statement the solicitor said, "Representations by any persons or organizations to the effect that they can, under existing law, obtain a refund of processing taxes for a farmer who was not a processor are false and misleading.... Likewise, representations by any persons or organizations to the effect that they can, under any future law, obtain a refund of processing taxes for a farmer, whether signer or non- signer, which the farmer himself could not obtain, are also false and misleading."[41]

Gurney wrote Mueller, saying that he had received a large number of copies of the release by the United States Department of Agriculture. He said he had answered the Agriculture Department statement on his radio noon hour program numerous times, as "they keep popping up from everywhere." He commented that the persons who wrote the statement "probably do not know that the Bill has already been before the investigating committee of the Department of Agriculture, and that the Sub Committee of the House has finished its investigation and recommended that it do pass, that Congressmen and Senators are falling over themselves now to get back of the Bill."[42]

Gurney told Mueller that it was not necessary for him to go to his county agent to get evidence of his hog sales. The evidence was not needed until the bill became law. Gurney suggested that Mueller and other farmers with similar problems sign a statement requesting that the county agent furnish copies of the original sales slips when the bill was passed.[43]

Then Gurney, in a rare moment of irritation, commented on the Department of Agriculture's actions: "The Department of Agriculture is supposed to be about the only Department of Government that the farmer might call his own, but in this particular case, they are using every effort to hinder him in every way from securing a refund of approximately $361,000,000 that was collected from him by an unconstitutional Bill."[44]

In addition to questions concerning the status of signers and nonsigners, and problems of locating records of their hog sales, farmers also told Gurney of their feelings concerning the New Deal farm programs. Many felt that the processing tax had been unfair. Hubert Pool of Delbert, Minnesota wrote, "Hope you will be able to collect this unjust tax for the farmers."[45]

Paul Klitzke of Elkton, South Dakota, hoped that Gurney would be able to "get this money back for the farmers that was taken from them."[46]

William H. Schaller of Groton, South Dakota, wrote that he understood Gurney was "helping the farmers get back their rightful money."[47]

C.E. Pickett of Cozad, Nebraska, referred to "my money which they took from me."[48]

C.H. Compton of Cambridge, Nebraska, wrote, "I have always contended this was an unfair tax on the producers."[49]

C.P. Knapp of Westfield, Iowa, commented, "They took about $1500 from me." [50]

M.A. Brady of Hay Springs, Nebraska, wrote, "It was just a steal."[51]

C.H.T. Jenson of Nelson, Minnesota, wondered if the government would have to keep back about 50 percent of the refunds for the "work of mailing out the money to us farmers. But they shure [sic] as hell should not do so as we farmers were not to blame they took that money. I fought that deal as hard as I could. I have not singed [sic] up on any of the foolish thing yet."[52]

Other farmers feared that they might lose the right to make their own decisions under the New Deal farm programs. Howard Abbott of Webster City, Iowa, said that he was a signer of the AAA program, although he was never much in favor of it. He feared if farmers kept on signing up they would "soon all be slaves for the government, lose our freedom."[53]

Elmer Rohm wrote, "We are willing to try and help get action on this as it sure made us angry, as lots of other measures has since. A sort of Communistic way of doing we think."[54]

F.J. Matezeek of David City, Nebraska, wrote that he and his family had not signed up with the AAA program, and "are still independent."[55]

In May and June of 1938 Gurney had good news to report. Writing to Carl Wall of Mellette, South Dakota, Gurney said the House and Senate subcommittees had approved the bill. The Senate Agricultural Committee had reported favorably on the bill, and he commented, "Of course, I am enthused over the progress that has been made."[56]

In a letter later that month Gurney remarked: "Since I have been working on this, considerable progress has been made. Understand please that the legislation has not yet been adopted that will make this refund payable to you, but we must continue to work for this if we are to expect that this illegal tax will be paid back...."[57]

Gurney also suggested to those who listened to him on the radio or those who wrote to enlist in his program that they write their senators and representatives, encouraging them to vote for the hog processing tax refund bill.[58]

In June Gurney traveled to Washington and met with various congressmen and Edward E. Kennedy, the lobbyist for the National Farmers Processing Tax Recovery Association. They discussed the progress of the hog processing tax refund legislation. On returning home, Gurney reported his findings in letters to farmers who were joining his program.[59]

In July Gurney was able to report to a correspondent that the delegates to the South Dakota Republican Convention had placed a demand for the return of the hog processing tax in their state platform.[60]

It is surprising that D.B. Gurney did not report in any of his letters a piece of family news, the news of his son, Chandler Gurney's, entry as the Republican candidate in the contest for the South Dakota United States Senate seat. According to news reports, Chandler Gurney criticized the New Deal, saying that it had failed to bring recovery and was possibly leading the nation into a "dead end."[61]

In November of 1938 Chandler Gurney was elected to the United States Senate.[62]

In the months following the election, D.B. Gurney continued his program to secure the passage of the refunding of the hog processing tax. He wrote letters reminding farmers who had sent in their claims but not paid their enrollment fees. Some farmers wrote back to explain why they could not pay their fees.

Gurney replied that he would try to enroll as many as he could, and he recognized that many farmers were having difficulties in financing even their initial fees. However, "the expenses are very heavy and I cannot afford to carry this load myself."[63]

By January of 1939 Gurney had about 800 paid-up members of his organization.[64]

According to Jan dePagter, who worked with Gurney during these years, Gurney "sincerely believed this cause was just and he put his whole heart into the fight."[65]

He may also have been motivated by a dislike of the Roosevelt Administration and the AAA. His first inclination was to especially help the farmers who had refused to sign up with the AAA. It was only later that he began to favor including both the non-signers and the signers in the processing tax refund bill. Gurney was a Republican, but a particular kind of Republican — a Republican who had supported Lemke, for years. Lemke in addition to being fervently interested in programs for improving the lot of farmers suffering from the problems of the Depression, was a vigorous critic of Franklin Roosevelt and the AAA from 1934 on.[66]

Although Gurney did not generally criticize the New Deal or the Department of Agriculture in his letters, sometimes he let slip a comment that revealed his dislike of the AAA. In one letter to a farmer in Iowa who wrote that most of the farmers in his area had signed up with the AAA, Gurney commented that he suspected not everyone had signed up in that particular township, and "I guess that many of them wished now that they never did sign up."[67]

How did the National Farmers Process Tax Recovery Association leaders feel about Gurney's activities? Gurney's campaign to help farmers recover their processing taxes had been regarded by the leaders of the National Farmers Process Tax Recovery Association with mixed emotions. They appreciated the publicity that his broadcasts had given to the cause, but they resented the competition. Association recruiters complained that farmers who might have joined the Recovery Association enrolled with Gurney instead. The Association leaders were uncertain as to whether to actively cooperate with or oppose Gurney's campaign.[68]

Then, as Chandler Gurney prepared to leave South Dakota to serve in the United States Senate, D.B. Gurney decided that his continued active participation in the campaign to recover the processing taxes might place Chandler Gurney in a position where he could be charged with conflict of interest. So D.B. Gurney decided to diminish his role in the campaign to recover the processing tax. He began by telling potential subscribers that they and their friends and relatives could enlist with him or with the National Farmers Process Tax Recovery Association. Either group could register a farmer's claims for the refund of the processing tax.[69]

At least Chandler Gurney's election was the official reason given for the change in tactics. There may have been other factors influencing D.B.

Gurney's decision. The response from farmers had not been as great as Gurney had hoped. His finances were in a low state. Radio station WNAX was facing a difficult struggle for re-licensing and was eventually sold in November of 1938. D.B. Gurney's health was also a concern. According to family tradition, it was D.B. Gurney who was first asked to run for the Senate, but he declined on the grounds of poor health.[70]

For whatever the reason, or combination of reasons, D.B. Gurney decided to take a less active role in the fight to recover the processing tax. He met with leaders of the National Farmers Process Tax Recovery Association and made arrangements to turn over his accounts to them. He sent his letters and files to the Association's offices. He no longer spoke over the radio concerning the processing tax nor wrote letters to farmers. All letters addressed to Gurney asking for information on the processing tax refund activities were forwarded to the Recovery Association offices in Des Moines and answered from there.[71]

For nearly a year after he withdrew from active participation in the program, Gurney paid $50 a month to Edward E. Kennedy, the Association's lobbyist, to continue lobbying for the return of the processing tax.

In the Senate, Chandler Gurney enthusiastically supported bills to refund the hog processing tax to farmers.[72]

The bills were discussed in the agricultural committees of the House and the Senate, but never passed by Congress. They were opposed by the Department of Agriculture, whose officials claimed that the farmers who cooperated with the AAA and signed the contracts had already received their reimbursement. Those who did not sign the AAA contracts had also benefited by AAA actions. They received the higher prices for their hogs because of the corn-hog adjustment program.[73]

Besides, the officials added, the cost of repayment would be too great.

The National Farmers Process Tax Recovery Association officials finally gave up the battle to recover the processing taxes in 1942. The records were saved, but no further action was taken after January or February of that year.[74]

The bills were not really supported by the national farm organizations.[75]

For a time, the Farmers Union officials supported the bills, and president of the Farmers Union, John Vesecky, spoke in favor of the bill before a House subcommittee.[76]

However, the Farmers Union was undergoing a leadership struggle. The leadership of the Farmers Union by 1940 was moving toward a position of support of the Roosevelt Administration. Vesecky did not remain president of the Farmers Union for long. Leaders opposed to Kennedy, and less supportive of the bills for the refunding of the hog processing tax, were elected. The American Farm Bureau, which was the largest farm organization, had enthusiastically supported the AAA in 1932 and never supported the hog processing tax refund bill. Some of the Farm Bureau leaders may have opposed it. At the last moment, the National Grange sent a letter to one of the hearings in support of the bill, but no representatives of the Grange appeared at the hearings.[77]

Later, the demands of World War II diverted national attention to other matters.

For a year D.B. Gurney had enthusiastically campaigned for the recovery of the processing tax to the farmers. It was a campaign that failed. The farmers who wrote to him never formed a strong organization that could rally support for, or influence enough legislators to secure legislation for, the return of the processing tax. Perhaps it was a fault inherent in the medium that he used. How can a radio audience be organized? Perhaps it was the strong opposition that the proposals faced. Or perhaps it was the problem of timing: Legislation which Gurney and the Recovery Association sponsored failed to pass in 1938; Gurney withdrew in late 1938 and the drive lost an important leader. In 1939 the bill was originally passed by both the House and Senate Agriculture Committees and then was recalled by the House Agriculture Committee and defeated by one vote. Following this second defeat, the drive to recover the processing tax began to lose momentum as the war clouds gathered over Europe and Asia, and national attention was focused in international directions.

7

"A Program for the Farmers": The Department of Agriculture

The Department of Agriculture had been established in 1862. Throughout its long years of service it had been an important but never particularly large organization. Originally, research had been its primary function. In 1933, because of the Depression, the Department developed and administered vast programs of agricultural adjustment, requiring thousands of new employees and a greatly expanded budget. With the many changes in size and function came questions and disagreements concerning agricultural programs.

While they wanted help with the severe problems besieging agriculture in the Depression, Farmers Union leaders had hoped that Roosevelt would support cost of production. They actively promoted his nomination and election in 1932. Although he had talked with some of the Farmers Union leaders, and seemed — before the election — to approve of their programs, Roosevelt came to support the ideas of men such as Rexford G. Tugwell and Raymond Moley who had different plans for the future of agriculture. Tugwell and Moley were both college professors, with little experience in agriculture, who believed agriculture must adapt to a primarily industrial nation.[1] Tugwell had met M.L. Wilson, a Montana agricultural economics professor, in Chicago in 1932 and brought back Wilson's proposals for a domestic allotment program. Roosevelt gathered advisors to help him formulate proposals for a farm program based on agricultural allotments. One of the people he invited to meet him and to help with formulating a farm program was Henry A. Wallace of Des Moines, Iowa. Wallace, the son of a former Secretary of Agriculture, was editor of *Wallace's*

Farmer, a popular farm newspaper in the Middle West, and also president of a hybrid seed company.

Following the election, Roosevelt vacationed in Warm Springs, Georgia, and considered possible cabinet appointees, including candidates for the position of Secretary of Agriculture. William Hirth of the Missouri Farmer's Association, John Simpson of the Farmer's Union, Henry Morganthau, George C. Peek and Henry A. Wallace were all possible candidates. Tugwell and Wilson worked hard to persuade Roosevelt to choose Henry A. Wallace, someone who favored their ideas of production control through a domestic allocation plan.

When Edward A. O'Neal, president of the American Farm Bureau, came out in favor of Wallace, this tipped the balance of the scales in his favor.[2]

As the new secretary of Agriculture, Wallace was determined to enact a domestic allotment plan. Wallace invited farm leaders to Washington for a National Agricultural Conference in March of 1933. There Wallace expected that the influence of an "inner group of moderates and progressives would prevail."[3] Those attending the National Agricultural Conference endorsed the domestic allotment plan in principle; and after the conference adjourned, a smaller group from the Department of Agriculture proceeded to write the Agricultural Adjustment bill.[4]

The only important farm leader who was not at the conference was John Simpson of the Farmers Union. Later he was to charge that he had been deliberately misled so that he would not attend.[5] In the Senate hearings on the Agricultural Adjustment bill, Simpson argued against domestic allotment and called instead for cost of production.[6] Although most farm groups, including the Farm Bureau and the Grange, supported Wallace's proposals, the Farmers Union, the Farm Holiday Association and the Missouri Farmers Association still called for cost of production.[7]

Wallace argued that it would be impossible to determine prices for agricultural products accurately under the cost of production plan. Wallace's argument, however, failed to convince Farmers Union critics who knew that cost of production figures for various crops were published in the Department of Agriculture 1930 Yearbook.[8]

While the debate was raging in Congress, many farmers wrote to express their opinions. A chicken farmer near Tyler, Texas, commented, "Simpson seems to have the only sensible conclusion there is to the farming subject.[9] A Nebraska farmer wrote to President Roosevelt: "Farmers of Western Nebraska ask you as their president to insist that the marketing bill carry John A. Simpson's cost of production amendment."[10] Another farmer suggested Wallace needed "one good horse-sense Iowa farmer on your right hand." He said that most of the farmers he knew favored government-fixed minimum prices, and he felt that "John Simpson is more right than some other advisers."[11] Fred Schultheiss, a member of the Farmers Union executive committee, wrote President Roosevelt: "I fear you must sift out the rubbish among your advisors and … listen to such leaders as John A. Simpson and Milo Reno, who represent the hope, the aspirations and the demands of at least 90 percent of the real dirt farmers."[12] A South Dakota farmer also wrote to President Roosevelt, "I plead with you to call in John A. Simpson and consider his relief measures."[13]

On March 11 and 12, Farmers Union officials meeting in Omaha encouraged their members to write Secretary Wallace and President Roosevelt in support of the cost of production. Many of the letters were similar, but others expressed more individual views on the subject. Agricultural historian Gilbert Fite comments, "The contents of hundreds of hand written letters indicate that farmers were thinking for themselves and were not pawns for any organized group."[14]

During the Senate hearings Simpson testified for three days, attacking the Administration bill as a subsidy to consumers because under it they could buy farm products at less than the cost of production. Simpson also said that the bill would not decrease production because farmers would cultivate their remaining acres more intensively, and big farmers and insurance companies would rent unused land and turn it into production.[15]

Simpson also attacked the Administration's plans for control of production because, he said, "You would have to have God on your side to be sure that such a program would work." The government's program, he felt, was "doomed to failure."[16] An Iowa farmer wrote to President Roosevelt: "To cut down production is to say that God does not know how much we need." He further commented that there could be no overproduction when millions were starving.[17]

Under the leadership of Senator George Norris, the Senate added a cost of production amendment to the Administration bill. Secretary Wallace sent a letter to the conference committee opposing cost of production as impossible to implement or enforce. Simpson appealed directly to President Roosevelt, reminding the President of his support during the elections. But President Roosevelt supported Secretary Wallace rather than Simpson, and the cost of production amendment was dropped from the final version of the Agricultural Adjustment Act.[18]

Other additions were made to the Agricultural Adjustment bill. One was an amendment proposed by Senator Elmer Thomas of Oklahoma, also supported by the Farmers Union. This amendment called for inflation to help farmers get out of debt. President Roosevelt conferred with Thomas and got him to moderate his proposal so that it became an instrument that the federal government could use for currency expansion if officials so desired. [19]

The Administration applied a great deal of pressure to Congress, and the bill passed. President Roosevelt signed the Agricultural Adjustment Act on May 12, 1933. And Wallace and Tugwell went to the White House to watch President Roosevelt sign what was, in effect, "their bill."[20]

According to the terms of the Agricultural Adjustment Act, seven major commodities were subject to control: cotton, corn/hogs, wheat, dairy products, tobacco and rice.[21] Because the Agricultural Adjustment Act was passed in May when cotton was already growing in the fields, cotton growers were paid to plow up part of their crop. This was the first crop to be reduced under the Agricultural Adjustment Act.[22]

Since corn and hog prices are closely related, the Department of Agriculture decided to deal with them together. There had been a large corn crop in 1932, and this caused the price of corn to drop below $.20 a bushel in December of 1932. The low price of corn had encouraged hog producers to keep more hogs for feeding, and the prices for hogs had also dropped.[23] Department officials decided that the best means of dealing with the hog surplus was to have an emergency pig/sow slaughter. The emergency pig slaughter reduced the number of hogs from 62 million at the end of 1932 to less than 59 million at the end of 1933. Farmers were also encouraged to sign contracts to reduce the number of hogs they produced.[24]

Both the farmers who brought pigs and sows to the emergency pig/sow slaughter, and those farmers who signed the corn-hog contracts, were compensated by funds derived from a tax levied on the processors. The processing taxes were to be collected by the Bureau of Internal Revenue under the direction of the Secretary of the Treasury.[25] The rate of the tax could be equal to the difference between the market price and the parity price, but because the administrators feared that a heavy tax would suddenly increase the price of pork and diminish the demand, a low processing tax was levied at first and then increased.[26] The processing tax levied for November of 1933 was $.50 per hundredweight (live weight). It was increased to $1.00 between December 1, 1933, and January 31, 1934; during February to $1.50; and on March 1, 1934, to $2.25.[27]

A corn processing tax of $.28 per bushel was also announced. There were many protests, and after a hearing the processing tax was reduced to $.05 per bushel. This meant that the processing tax on corn could not fully fund the crop reduction payments. It was decided to treat the corn-hog program as a unit and use hog taxes to help support the corn reduction.

Also, AAA officials decided to fund the corn-hog reduction program by asking the Treasury to make advances against future hog processing taxes.[28]

Milo Reno and Edward E. Kennedy cooperated during the initial phases of the Agricultural Adjustment program until it became apparent that a central feature of the program for the corn-hog farmers of the Middle West was to be a hog processing tax. At that point they realized that the tax would probably be taken out of the price the packers paid the farmers, and only the farmers who joined the program would benefit. Kennedy and Reno believed "there was no way this kind of program would help the farmer." And they decided to oppose the Agricultural Adjustment program and its execution by the Department of Agriculture.[29]

In order to reduce production, the Agricultural Adjustment Administration called for a reduction in acreage of corn and other commodities, and in the numbers of hogs. The reduction was planned first on a national basis and was designed to provide a supply sufficient for the domestic market. It was also planned to provide some products for export markets (which had been decreasing). This allotment was then apportioned among the states and counties according to their past production records. Each cooperating farmer's allotment was worked out by local and county committees. The farmers who cooperated then received compensation for their reductions.[30]

During the years of the Agricultural Adjustment Administration, 1933–1936, the program received much criticism. Much of the criticism focused on the slaughter of pigs and the lower hog prices, which farmers blamed on the processing taxes. The loudest voices of opposition were raised by members and leaders of the Farmers Union. Critics of the slaughter program said that the slaughter was wrong, and a better program would be to distribute free pork products to hungry people. Letters came from pork processors, farmers and unemployed workers saying that poor people in the United States did not have enough to eat, and the Department policies should be changed from production control to programs to feed the hungry. Speaking before the Senate committee on the AAA bill, Simpson had predicted this when he testified: "The farmers are not producing too much. We need all this. What we have overproduction of is empty stomachs and bare backs."[31]

This same concern for the effects of the Administration policy of reduction of surpluses on starving people was expressed by William Hirth, leader of the Missouri Farmers Association, who commented in a letter to James Farley, Democratic Party National Chairman, that there should be no efforts to reduce surpluses until all of the 120 million Americans were eating three square meals a day.[32]

Department officials replied to these criticisms. S.W. Lund, Assistant to the Chief, Meat Processing Section, in a letter to a pork processor, wrote that the original emergency hog slaughter plan had been expanded. "The original plan as to the number of pigs to be slaughtered has been enlarged, especially to make provisions for the marketing of pigs from drought areas. You will be further interested to know that plans are well advanced to distribute meat now being saved and processed to destitute families in every State. This will be furnished to them in addition to the meat they are now getting through local relief agencies."[33]

But the issue would not go away. In June of 1935 the *National Union Farmer* carried an article on J.H. Saucke of Farnhamville, Iowa, who said he killed 42 pigs so that he could be in compliance with his corn-hog contract. The article, accompanied by a picture of the dead pigs, was on the front page of *The National Union Farmer*.[34] Department officials said they had requested the Iowa State Corn-Hog Committee to investigate the matter. Apparently Saucke had bought some extra sows who farrowed 42 little pigs, which placed him out of compliance. Saucke had then shown the local corn-hog officials a statement from a rendering plant employee saying that he had delivered the extra pigs to the plant; but the local committee members questioning this account looked into the matter more thoroughly. Saucke then became panicky and had a representative of the rendering plant kill the pigs. Concluding his report, the Department official stated, "It is doubtful if further publicity in this matter upon the part of the Agricultural Adjustment Administration would do more than to give this unfortunate case further notoriety, which perhaps would be welcomed by those who are in opposition to the activities and purposes of the Agricultural Adjustment Administration."[35]

The case received considerable publicity. Reno mentioned it in a speech, and news articles concerning the Saucke case were forwarded to President Roosevelt, who requested a report from the Department of Agriculture. Tugwell wrote the President and also enclosed an interdepartmental memorandum that concluded that it was unfortunate that Mr. Saucke chose to kill his excess pigs rather than to turn them over to the Federal Emergency Relief Administration for distribution to relief. His action prevented some needy people from receiving surplus pork, and it also gave Reno the "opportunity of publicizing several distorted facts concerning the incident."[36]

In December of 1935 Secretary Wallace spoke over the radio program *National Farm and Home Hour* about the letters he had received concerning the killing of little pigs. He said that the pigs were mentioned more often than any other subject except potatoes in the letters he received. One letter writer said that it made him sick "all over" to think of how the government had killed millions of little pigs and how that had led to the increased prices of pork. Secretary Wallace said that it was common belief that the increased prices of pork were due to the 1933 pig slaughter. But that was not true. There was more pork in 1935 at lower prices because of the emergency pig slaughter. Wallace said that he was used to statements of this type from politicians, demagogues, newspaper writers and others who were driven by their partisan beliefs so that they could not distinguish the truth. But the truth was, as he saw it, that the production control of the AAA had been necessitated by the conditions of the times. Wallace continued: "I suppose it is a marvelous tribute to the humanitarian instincts of the American people that they sympathize more with little pigs which are killed than with full grown hogs. Some people may object to killing pigs at any age. Perhaps they think that farmers should run a sort of old folks home for hogs and keep them around indefinitely as barnyard pets." Still, he believed that it was more important to think about farmers and consumers than about hogs.[37]

Other criticism of the corn-hog program dealt with the processing tax. Various groups protested the tax, but Farmers Union leaders were some of the most strident critics. In a radio address Reno declared, "The highway robber, who throws a gun on his victim and gives him the choice of delivering up his cash or taking the consequences, is allowing his victim the same choice of voluntary cooperation as the Triple A when it says to the farmer: Sign this contract; deliver, ... or we will not only take $2.25 upon every hundred pounds of pork you produce, but we will boycott and harass you in every way possible.... In fact, unless you recognize the power of this six-shooter we hold in our hand, you will eventually be destroyed."[38]

Many farmers complained to their congressmen that since the processing tax was applied they had received lower prices for hogs. A Georgia farmer complained to Congressman Homer Parker that some hogs in his neighborhood were selling for $.02 a pound. Parker wrote to George Peek demanding that he stop the packers' reducing the prices they paid for hogs. Fearing that if practices such as these were allowed to continue they would lower farm prices, he urged that the Department of Agriculture take "the necessary steps" to stop these unfair practices.[39]

A farmer in Michigan complained to his congressman, M.J. Hart, concerning the tax. The congressman wrote to Secretary Wallace: "I am hoping that the Department would not find it necessary to increase [the tax] because I am fearful that it would drive the price down just that much more...."

Secretary Wallace then wrote a letter to Congressman Hart in which he stated that it was "difficult to make a definite statement as to which class is bearing the cost of the tax." But he believed that there had been a "misinterpretation by some farmers who feel that the entire tax has been taken off producers' prices." The effects of the tax had been made to look larger, he felt, because they came at a time when hog prices usually decline. Furthermore, hog producers should remember that the processing taxes provided the revenue for the AAA production control payments. "In other words, a large percentage of the money collected is actually returned to the producers who participate in the corn/hog program."[40] Thus, while Wallace was denying the idea that the whole processing tax was being borne by the farmers, he seemed at this time to be admitting that some part of it was borne by them.

Farmers also petitioned the government to change the application of the processing tax on hogs. Farmers and hogs producers of DeKalb County, Missouri, charged:

"That the packers are deliberately taking the processing tax off live hogs, which is contrary to our understanding of the government's application of the processing tax.

We further respectfully ask that you apply all the power granted to you by the A.A.A. and the Packers Stockyard Act or any other powers at your command to correct this injustice."[41]

Cecil A. Johnson, executive assistant in the corn-hog section of the AAA, wrote to a Missouri farmer, "It is regrettable that there are those who have formulated opinions to the extent that they believe the processing tax is responsible for the present price of hogs.... We agree that this price is unsatisfactory and [therefore we are trying to] carry out the corn-hog program." He said that the processing taxes were necessary because the AAA had borrowed on future processing taxes to make payments to cooperating farmers.

Johnson and M.L. Wilson said they did not believe the farmers were receiving less because of the tax; instead, farmers would receive more for their hog production than if no program had been undertaken.[42] In regards to the farmer who did not sign up for the program, Wilson said, "It has been our aim to develop a program whereby the non-cooperator would receive as great a return for his hogs as he would if there were no program, but at the same time reserving the major share of the benefits to those who actively participate in the program. We firmly believe this objective is being attained."[43]

By late 1935 there was a new complaint about the corn-hog program and the processing tax. Prices of pork and other products had risen by 1935. Letters and petitions were sent by meat retailers and consumers requesting that the processing tax on pork be abolished because it raised the price of pork. USDA officials stated that the slaughter of little pigs in 1933 did not lead to higher prices in 1935. The pigs slaughtered in 1933 would probably have come to market and been eaten in 1934. The sows slaughtered in 1933 would have produced little pigs which would have come to market by the end of 1934. The AAA program with normal crop yields would have produced ample meat supplies for 1935. The drought caused the meat shortage, not the AAA.[44]

The effects of the processing tax were discussed within the Department of Agriculture by various members of the Department. Louis Bean, Economic Adviser to Secretary Wallace, attributed lower hog prices to a seasonal decline between September and December. In the fall of 1933 the decline was greater because of the large volume of sales, caused in turn by higher corn prices, which made keeping hogs seem less desirable. "Possibly," wrote Bean in a memorandum, "the propaganda by interested parties that the effect of the Federal program would be to bring about lower hog prices may also have induced farmers to market more heavily than usual." The lower prices received by competing meat products may have contributed to the lowering of hog prices. Also, consumers' incomes did not increase in the winter of 1933–1934, and this may have held down prices. "These considerations," Bean continued, "lead me to the present tentative conclusion that the effect of the processing tax on hogs has worked in both directions. The extent to which the tax may have lowered the farm prices, has, of course, a different significance to the participating farmer than to the non-participator. The former will get a benefit payment; the latter has little to gain from the operations of a processing tax on hogs." He concluded, "It might be well ... to point out to the farmers ... that it is to their advantage to join the program rather than to stay out of it."[45]

In another interdepartmental memorandum, Alfred Stedman, head of the information office of AAA, wrote that it was essential to have a study of who—farmers, processors or consumers—paid the processing tax on hogs. The matter was urgent. "Indications are that this will be the first inquiry which we may be called upon to answer when discussion of the Agricultural Adjustment Administration begins in Congress." He said that Representative Marvin Jones and Senator Pat Harrison both felt that the Department of Agriculture should be prepared to discuss this point. Stedman was concerned about the possible adverse effect the processing tax was having on agricultural opinion in the Middle West, and he offered to prepare a press release when the studies had been completed.[46]

An interesting phrase often used in these discussions was "propaganda regarding the hog tax." Department of Agriculture officials felt that the processors were contributing to this propaganda and wondered if they were encouraging Farmers Union leaders. In late 1934 and early 1935 Milo Reno attacked the AAA over the radio nearly every Sunday afternoon. Secretary Wallace wanted to know who was paying for Reno's speeches.[47]

In answer to criticism, the Department held two corn-hog referenda in 1934 and in late 1935. In both of these referenda few non-contract signers voted. In 1934 approximately 374,000 farmers voted to continue the program; approximately 161,000 voted against the program. Only 45,000 non-signers indicated their preferences on agriculture ballots.[48] The Department of Agriculture took these votes to mean that farmers favored the continuance of their program. Another referendum was held in 1935 with similar results. The questions were: (1)"Do you favor an adjustment program dealing with corn and hogs in 1935?" (2)"Do you favor a one-contract-per-farm adjustment program dealing with grains and livestock to become effective in 1936?" In 1934 approximately 46 percent of all contract signers voted in favor of the program.[49] In 1935 the question of the referendum was: "Do you favor a Corn-Hog Adjustment Program to follow the 1935 program which expires November 30, 1935?"[50]

Articles in the *National Union Farmer* indicate that some Farmers Union officials felt the 1935 referendum was fraudulent. They said that farmers were asked to vote on a "trick-question, so worded by the bureaucrat chiefs that the answer could hardly be other than favorable." Some Farmers Union officials charged that marked ballots were used in Michigan. "While pretending to conduct a secret ballot, the men at bureaucrat headquarters had covertly inserted serial numbers of all signers so that they could quickly check back to the contract form and discover in each case, how the farmer voted or whether he voted at all." Since these officials controlled the handling of contracts, farmers could be intimidated into voting to approve the program. Coercion was also charged in the AAA referendum in Iowa.[51] In Iowa AAA fieldmen were paid to go out and secure ballots from the farmers and take these ballots to the polls. Only those who gave their votes to the AAA fieldmen were counted, since the Iowa polling places were closed on election day.[52]

Approximately 50 packing companies challenged the processing tax in court.[53] The cases moved slowly, and many large corporations withheld the disputed taxes in special accounts. During 1935, while AAA payments to corn and hog producers were approximately $330,000,000, collections amounted to approximately $191,500,000. There were several reasons for the shortfall. One of these was the drought of 1934, which had affected the numbers of hogs farmers raised; but one of the most important was the refusal of the processing companies to pay their taxes pending court decisions.[54]

On January 6, 1936, the Supreme Court, in *U.S. v. Butler,* held the AAA to be unconstitutional.[55] When the news of the Supreme Court decision reached the offices of the AAA in Washington, there was extreme confusion. Approximately $200,000,000 in processing taxes had been impounded, and AAA officials feared that the money would be returned to the processors. The AAA owed $217,250,000 to farmers for their compliance with crop reduction programs in 1935. Some officials estimated that $200,000,000 more than had been taken in from processing tax revenues had already been distributed.[56] Where was the money to make promised payments to come from? Secretary Wallace was silent during those first few days and said he would see no newspapermen.[57]

President Roosevelt conferred with leading AAA officials and chairmen of the Senate and House Agriculture committees, and made plans to introduce a bill immediately for the appropriation of $250,000,000 for payment of the farmers' claims.[58]

Following the decision, the Department of Agriculture called for a meeting of farm organization leaders to make plans for new legislation. Secretary Wallace said he "wanted to get the counsel of the wise leaders of agriculture."[59]

Some farm leaders already had expressed their feelings. Edward A. O'Neal, head of the American Farm Bureau Federation, called the ruling "a stunning blow to national economic recovery." O'Neal said: "Those who believe the American farmer is going to stand idly by and watch his program for economic equality and parity, for which he has fought more than a decade, swept into discard, will be badly mistaken."[60] Louis J. Tabor, Master of the National Grange, commented that the administration needed to find a way to fulfill the contracts made with the farmers under the agricultural adjustment program.[61] Milo Reno declared: "That's fine. It was unthinkable legislation in the first place."[62]

In Washington the group met with Wallace and then went into executive session. Although Milo Reno had not been invited, the leaders heard that Reno intended to "crash the gates." Eventually, Reno attended the session as a member of the Farmers Union delegation. The Farmers Union leaders called for cost of production, and licensing processors and distributors to guarantee against the sale of commodities at less than cost.[63] Officials of the National Grange proposed a plan calling for increased agricultural tariffs, subsidized exports and cooperative marketing.[64]

The Department of Agriculture officials had said they called the conference to get ideas for a new approach to agricultural problems. There is some evidence indicating, however, that from the first they favored legislation calling for payments to farmers who practiced conservation. On

January 10, the day of the conference, President Roosevelt said, "We shall try to get some legislation at this session which will carry out in some way the general thought of seeking to maintain ... soil fertility because we have lost an awful lot of it and, at the same time, keep the price for American agricultural crops up to a high level."[65]

Howard R. Tolley, a former assistant AAA administrator, was called from the West Coast to attend the meetings. He had been developing proposals for regional planning and soil conservation for the Ginanini Foundation of California. Tolley was sometimes called the "Soil Wizard."[66] The invitation to him was considered especially important since President Roosevelt, Secretary Wallace and AAA administrator Chester Davis had said they favored some sort of soil conservation program.[67]

The American Farm Bureau had called its officers and general board together in Washington to meet a day before Wallace's conference. They also discussed possible solutions to farm problems in the wake of the Supreme Court decision. Donald Kirkpatrick, American Farm Bureau general counsel, liked a proposal which may have originated in Chester Davis' office or may have come from recommendations of an Iowa farm study committee.[68] This proposal would be based on the welfare clause of the constitution rather than on the commerce clause, which the Supreme Court decision seemed to suggest could not be used. Using the welfare clause, the government would pay farmers to restore and maintain soil fertility through growing soil-building crops to be plowed under rather than harvested. Kirkpatrick believed this could result in a reduction in harvested acres and therefore in lowered agricultural production.[69] Kirkpatrick passed his suggestions on to Ed O'Neal and Earl Smith, president and vice president of the American Farm Bureau, and they spent several days with AAA administrators discussing the new proposal.[70]

At the farm leaders conference they appointed a committee of 13 to draft proposals for a new program to be submitted to the group.[71] The committee hammered out a declaration of principles that they felt new agricultural legislation should include: 1. The Secretary of Agriculture should be given the power to withdraw from commercial crop production such land as might be necessary to promote conservation of the soil and to bring about a "profitable balance of domestic production with the total effective demand at profitable prices." 2. Congress should provide adequate funds for carrying out such a program. 3. Processing taxes might be used if voted on by a reasonable number of producers, approved by processors, and used to open up new markets. 4. All valid provisions of the old AAA should be retained. 5. Marketing of farm products should be done as far as possible through farm cooperatives. 6. The American market should be preserved for the American farmer, and the expansion of foreign markets should be sought for American surplus crops.[72]

At the end of the conference the general statement was endorsed by agricultural representatives. News announcements released by the Agricultural Department declared that a general statement had been unanimously approved by the nation's farm leaders.[73]

Writing in the Farmers Union paper, Edward E. Kennedy, a Farmers Union representative at the conference, commented that the general statement of the conference was "not entirely clear and not thoroughly satisfactory to the Farmers' Union. However, it does recognize the principle of genuine equality for Agriculture. It recognizes the further fact that whatever is done must be done to secure 'profitable prices' to the farmers." Kennedy also believed the conference had recommended that the farm organizations should prepare legislation and present it to Congress.[74]

After the farm representatives meeting, a two-year Soil Conservation and Domestic Allotment Act was introduced into Congress.[75] Many of the most important provisions of the Soil Conservation Act were passed by Congress within eight weeks after the Supreme Court decision.[76] The Administration plan called for payments to farmers who agreed to reduce production of soil-depleting crops and maintain erosion preventing and soil-improving crops.

The legislation was supported by groups connected to the AAA. The Farm Bureau, tied closely to the AAA through its special relationship to county agricultural agents, and the Farmers National Grain Corporation, with financial ties to the federal government, approved the Administration legislation of payments to farmers who agreed to practice conservation. Farmers Union leaders were disappointed because they had hoped for cost of production legislation, and Grange leaders had wanted cooperative marketing, export subsidies and increased agricultural tariffs. The text of the Soil Conservation Act was reprinted in the Farmers Union paper, with an introductory note that the law was written by "unknown persons in the Department of Agriculture" and considered in executive sessions of the Senate and House Agriculture committees. The writer of the note commented: "The so-called Wallace conference held in Washington January 10th and 11th might just as well not have been held."[77]

Following the passage of the Administration bill, Reno returned to Iowa. He was angry and disappointed that Farmers Union ideas of cost of production had not been enacted. He soon proposed the organization of a farmers group to recover the money that he said had been deducted by the processors in the prices they paid the farmers for their hogs. The farmers he attracted were often members of the Farmers Holiday and Farmers

South Dakota farmer engaged in soil conservation. (South Dakota State Historical Society Archives, Pierre, S.D.)

Union. They had generally been opposed to the first AAA, and many had not signed up for its crop reduction programs. They believed that their hog prices had been reduced by the packers, that they had paid the tax; they had not received any payments under the AAA, and now they wanted their money back. They first called their organization the Farmers Process Tax Recovery Association. Later, as they expanded, they called themselves the National Farmers Process Tax Recovery Association.

Meanwhile, the processors had organized to get the processing taxes back. First the processors demanded to keep the money they had set aside. Following the Butler decision, the Supreme Court unanimously returned the $200,000,000 that the processors had put into special accounts while they were testing the constitutionality of the processing taxes. Many people, however, opposed allowing the processors to keep the money.[78] Secretary Wallace supported confiscatory taxes on the processors. He openly questioned the justice of the Supreme Court's decision to return the impounded funds to the processors, arguing that the money in "most cases had already been passed on to consumers or back to farmers."[79]

In support of this position, Wallace requested information from his department advisors as to the composition of the processing taxes. Using Internal Revenue estimates of impounded processing taxes, Louis H. Bean, one of his economic advisors, reported that hog processing taxes accounted for approximately $51,000,000, wheat processing taxes for $67,000,000, and cotton for $51,000,000. [80]

The day after he received Bean's memorandum, Wallace met with President Roosevelt.[81] In March President Roosevelt sent a message to Congress suggesting a tax on the "windfall income" that the packers had withheld.[82] Congress passed a Revenue Act returning 80 percent of these monies.[83]

Because of the continued agitation against the processing taxes, and also to counteract the demands of the processors for return of the taxes, officials in the Department of Agriculture had requested that the Bureau of Agricultural Economics make a study of the processing tax. In response to the requests from the Department officials for information on the tax, the Bureau prepared *An Analysis of the Effects of the Processing Taxes Levied Under the Agricultural Adjustment Act.* Published in 1937 by the Bureau of Internal Revenue, the *Analysis* would be useful in fighting off the packers' claims, but it opened the door for claims by hog farmers. It was the culmination of ongoing studies that had been done since 1934.[84]

Writers of the *Analysis* discussed whether the processing taxes were absorbed by consumers, processors and distributors, or paid by producers through reduced prices.[85] They concluded that a very large part of the tax on wheat, rye and cotton was passed on to the consumer, but not in the case of hogs. This conclusion was based upon a study of the retail prices of commodities over a period of years, including the period of the processing tax.[86]

In evaluating whether the processors paid the tax, the economists looked at the spread during the period before the processing tax went into effect. From November of 1931 to October of 1933, the two year period before the levying of the processing tax, the spread between the price of hogs per 100 pounds and wholesale value of 71 pounds of pork products at Chicago was $.65. When the processing tax went into effect the spread widened sufficiently to allow packers to pay the tax and leave a balance about equal to the pre-tax spread. When the processing tax was increased, the spread again widened. Thus, in charts the economists showed that the spread for the periods from 1930 to June 1935 were $.65, $.70, $.67 and $.68.[87] After the tax was declared unconstitutional by the United States Supreme Court on January 6, 1936, the margin between the price of hogs and the wholesale value declined, and by the end of February was back to about $.70.[88]

The retail profit — that is, the difference between the wholesale and retail prices of the principal hog products at New York — averaged $2.35 for 52.6 pounds of pork products in the period from January 1930 to September 1933. From November 1933 to December 1935, when the processing tax was in effect, the margin averaged $2.36.[89] Since neither the processors nor the retailers paid the processing tax, the economists reasoned the incidence of the processing tax was almost entirely upon hog producers.[90]

Similar findings had been presented by Geoffrey Shepherd of Iowa State College at a meeting of the American Farm Economic Association in Chicago in December of 1934. Using data published in weekly bulletins put out by the Bureau of Agricultural Economics, Shepherd concluded that the packers were not paying the tax, the retailers were not paying the tax, and the consumers were not paying the tax. Thus, by process of elimination, he concluded that the farmers were paying the tax.[91]

The authors of the Bureau of Agricultural Economics *Analysis* stated that because the funds derived from the processing taxes were used to make benefit payments, the total income for cooperating producers (prices plus benefit payments) was "approximately the same as it would have been, under the prevailing conditions of supply and demand, if the tax had not been imposed."[92] This indicates that the AAA program did not substantially increase the income of farmers who belonged to the program. Those who were not in the program suffered a loss.

By 1937 the small group of Iowa farmers who hoped to recover their hog processing taxes had grown. Calling themselves the National Farmers Process Tax Recovery Association, they had engaged former national secretary of the Farmers Union Edward E. Kennedy to lobby for them in Washington. Kennedy was delighted by the publication of the Bureau of Agricultural Economics report. As discussed in Chapter 5, he used the *Analysis* to bolster the Recovery Association's claim that the taxes rightfully belonged to the farmers. He also used it as a recruiting instrument, sending copies of the BAE report to Donald Van Vleet, President of the National Farmers Processing Tax Recovery Association, and to state managers of the Association.

By January of 1938 D.B. Gurney joined the fight against the USDA criticism, labeling it as "propaganda," "absolutely false," and telling his listeners to "pay no attention to [it]."[93] On the air he denounced the articles as "misleading, false, malicious, libelous."[94] When discussing an article in the *St. Paul Farmer*, Gurney declared that he could not see "why a so called farm paper would be so bitter against the farmer."[95]

The Department campaign sowed a large seed of doubt in the minds of many farmers. Both Gurney and the Recovery Association found it increasingly difficult to get new members. Listeners wrote in that they were concerned about enlisting with Gurney in his program because they were hearing discouraging remarks about the program. An Iowa farmer confided that "the people in this country said it's the bunk, that there is nothing to it."[96] Another farmer wrote that he hoped Gurney would be successful, "although quite a few think that you are out to make some nice money."[97] One letter writer gave Gurney his neighbor's name and address and suggested that Gurney write the farmer and "give him a piece of your mind as he is telling around that you are trying to get rich on the campaign and we understand better because we hear you every day and he has no radio."[98] A Minnesota farmer wrote, "There is a great many people here that say there is nothing to getting the tax back."[99] Other listeners wrote Gurney about their discouragement. One wrote, "I feel it is not much use to do anything until it is passed, since there is a possible chance of it not being passed."[100] Another writer asked, "Mr. Gurney, do you honestly believe we will get the money? We surely do need it."[101] In June of 1941 a listener wrote, "Was you sincere in your promise to try and collect this money or was it just a money scheme to get some extra cash?"[102] A farmer from Madison Lake, Minnesota, wrote, "I don't care to send any money as I don't know what I would get out of it. And they're [sic] is too much red tape to all this. As the farmer always get [sic] the small end of every thing."[103]

The activities of the Department of Agriculture must have been directed against D.B. Gurney too. Gurney was reassured by A.J. Johnson, who commented, "I do not think you have anything to fear on the part of the government. They at one time not only checked upon our Association but they also sent a questionnaire to every individual who had filed a claim. However, nothing has materialized up to this time. I think it was done for the purpose of getting the farmers suspicious and also to put fear into the men who were in the field soliciting."[104]

In another letter to a farmer who had joined their program and sounded discouraged, Johnson wrote: "Your letter indicates that you are somewhat skeptical as to the possibilities of our organization ever getting this refund. I realize this is possible with all the propaganda that has been spread against our Association and also Mr. Gurney, whom you mentioned in your letter." But Johnson assured the farmer that the processing tax legislation had been passed unanimously by the Senate. He hoped the House would act on the matter in the early part of the next session.[105]

The final tactic that the Agriculture Department officials used against the Recovery Association was to oppose the bills that the Association had introduced in Congress. Some of the legislative maneuvers will be discussed in the following chapter. As the bills for the processing tax return entered Congress they met with some successes, but in the end always with failure. One of the decisive factors in their failure was the opposition of officials of the Department of Agriculture.

It had been a long struggle between the National Farmers Process Tax Recovery Association and the Department of Agriculture and the Agricultural Adjustment Administration; and in the end, the Administration won. The bills were defeated; the organization faded out of existence. The records were stored in A.J. Johnson's back bedroom to appear many years later in the special collections of Iowa State University.

What are some conclusions that can be drawn from a study of this conflict between the Department of Agriculture and the National Farmers Process Tax Recovery Association?

The Agricultural Adjustment Administration was established in 1933 as an emergency program, but the Department of Agriculture tended to look upon it as permanent, as something that must be protected from criticism.

The Department of Agriculture continually boasted that its programs were democratic, that they were formulated and administered by the farmers, but in the formation of the 1933 Agricultural Adjustment Act, and in the formation of the 1936 Soil Conservation Act, farmers were consulted after the Department generally had decided what the program would be.

The Department of Agriculture attempted to subvert any farm group that attacked its policies or could embarrass its programs.

The 1934 drought did more to raise prices of hogs than the AAA programs. This was admitted by many of economists when people in 1935 complained about high prices of pork products. Yet in 1938, when farmers wanted their money back, the officials said that the farmers had benefited by the AAA program, which raised prices.

Agricultural Department officials were inconsistent when talking about the effects of the processing tax. Ezekiel said the farmers were paying the processing tax. The Bureau of Agricultural Economics and others said farmers paid the processing tax, but those who signed up with the program received it back in benefits. Bean and Wilson said that the farmers who were not in the program were suffering a loss, but Wallace said farmers not in the program were not entitled to refunds because they had not paid the tax and had gotten higher prices as a result of the AAA actions.

The Department of Agriculture opposed the National Farmers Process Tax Recovery Association in several ways. Notices were sent to newspapers with the implication that the recruiting efforts were fraudulent. County agents were told not to assist the farmers in gathering records to prove their claims. Letters were sent to stockyards saying that efforts of farmers to get their records to prove their claims were foolish. The Post Office was asked to check up on the group. Johnson mentioned that the farmers who had filed claims also got letters from the government, which he felt were meant to frighten the farmers and perhaps keep others from joining the organization. Organizers and recruiters of the group felt that they were being singled out for special government attention. Some must have quit because of this.

Why did the Department of Agriculture fight so hard to oppose the National Farmers Process Tax Recovery Association? Perhaps all the criticism of the emergency pig/sow slaughter had made them extra defensive of the program. Milo Reno and William Hirth had both spoken out publicly against the emergency pig/sow slaughter.

It could have been anger at their persistent critics. Some of the leaders of the National Farmers Processing Tax Recovery Association — Reno, Lemke and Kennedy — had been leaders in the Farmers Union. They and Simpson, president of the Farmers Union until his death in 1934, had opposed the programs of the Department of Agriculture from 1933 on. Perhaps the Department felt that this was a way of retaliation. Certainly all that opposition would not have made the Department look favorably on any group which was formed and advised by those who might be considered enemies of the Department.

Wallace opposed the passage of refunds on the hog processing tax, but at the same time did not oppose the passage of refunds to cotton and tobacco producers. It may be that he permitted refunds to these other groups because they had the support of southern senators and representatives in Congress.

And there may have been other factors. Since the Recovery Association originated in Wallace's home state among farmers with whom he dealt in producing and selling hybrid corn seed for the corn-hog market, as well as people who read his farm magazine, hurt pride may have been a factor. Wallace may have resented the fact that some people from his home region rejected his program.

Another factor may have been connected with the success of hybrid seed corn. Wallace was one of the first to pioneer in the commercial production and sale of hybrid seed corn. During the last part of the 1930s the sale and use of hybrid seed corn rocketed, leading to increased production of corn.[106] This increased production of corn led to increased production of hogs. Yet he blamed the Association members who refused to join the AAA for increasing production, which was against the AAA policy of controlling it. Possibly Wallace was frustrated with his double roles of Agriculture Secretary responsible for a program of limiting production and his role as chief developer of commercially used hybrid seed corn (and chief stockholder of Pioneer Seed Corn, which was a very important factor in increasing corn production), and then transferred his frustration to the Association.

Some historians have noted that Wallace was surrounded by a small circle of "kingmakers," men who wanted to make him president. Perhaps these men felt refunding the farmers of the Middle West would strengthen the opposition, and that in the years before the election of 1940 the leaders of the Middle West must not be allowed to win even small battles.

A more clear cut reason for the Department of Agriculture to oppose the refunding of the processing tax to farmers was because the money had either not been collected or had been spent. The processors were, after all, allowed to keep 20 percent of the processing tax without questions. After that they could keep whatever they could prove in court they had not passed on to the consumers or producers. Much had been spent in paying back the farmers who had signed up with the program. Other funds were spent in administering the program, paying for the work of the committee persons, etc.

8

Edward E. Kennedy
and the Farmers Guild

The Farmers Guild was started by Edward E. Kennedy, an Iowa farmer who first served in the Iowa Farmers Union as a member, was active in the Farm Holiday Association, advised the Farmers Process Tax Recovery Association, and then organized the Farmers Guild.

He first came to the notice of the organization when he was asked by Milo Reno to come up with a formula for establishing the cost of production. According to Kennedy's recollection, Reno told him, "You are young, you seem to have imagination, and there are lots of things you don't know couldn't be done."[1] After asking the experts at Iowa State for help on establishing estimates of cost of production for various major farm products, and being told that it could not be done, Kennedy borrowed the formula used by manufacturers, and the figures supplied by the United States Department of Agriculture and other departments, and came up with figures for the cost of production. After his figures were accepted by the Farmers Union and other farm organizations, Kennedy sought to present them to President Coolidge who was vacationing in the Black Hills of South Dakota. However, this was a waste of time because soon after the interview, Coolidge announced that he did not plan to run again.[2]

In 1926 Kennedy left his Iowa farm and accepted the assignment of organizing the Illinois Farmers Union, moving to Kankakee in eastern Illinois. In 1931 he became Secretary of the National Farmers Union and continued to operate from Kankakee.

Also in the early thirties, Kennedy became active in the Farm Holiday Movement. Illinois farmers were attracted by the promises of the Farm Holiday movement. One observer wrote, "There is a revolt movement sweeping like wildfire over the corn belt states.... They are led by demagogues.... There was an immense meeting here last night. It was addressed by an organizer from Iowa [Glen Miller, president of the Iowa Farmers Union], and in his speech he blamed Herbert Hoover for everything which has gone wrong, from the bank failures to the corn borer."[3]

Farm Holiday members participated in rallies and protests to halt mortgage foreclosure sales in Illinois also. Near Carthage a crowd of 400 farmers gathered outside the Hancock county courthouse to protest the announced sale of a farm which had been held by a local farmer for forty-six years. Fred Huls, chairman of the Carthage Farmers Union, called the farmers together. Huls and other members of a committee appealed to Washington for a government loan to the farmer, and also achieved a grant of ninety days additional time for the farmer while the loan was being processed.[4]

There was an active unit around Kankakee.[5] They led in stopping foreclosures and holding penny auctions. They also worked toward electing judges who were sympathetic to the farmers.[6] When the Farm Holiday strike was announced, they participated. Fred Winterroth announced that the Kankakee, Will, Grundy and Iroquois county grain elevators and produce stations were closed due to the Farm Holiday strike.[7]

In the election of 1932 Kennedy seems to have supported Roosevelt. Illinois farmers and city dwellers alike voted Democratic in 1932, hoping for a change in the grim economic situation they were facing. The agricultural counties voted Democratic, believing that Franklin Roosevelt and his supporters would bring relief to the American farmer.[8]

Farmers began to write to complain about the AAA program. One farmer said, "If this continues, the corn-hog program will starve us all to death. [It is an] injustice as the fellows who raised too many hogs are getting the big bonuses while the little fellow[s] who went along in a quiet way are denied contract."[9] A widowed farm woman wrote President Roosevelt, "You talk help the Farmers, it's only the big ones you are helping."[10]

Despite their objections, most Illinois farmers followed the urging of the county agent and the county Farm Bureau and signed up for crop reduction contracts with the AAA.[11] In Illinois there were about 231,000 farms in 1934 and 1935. Of these farms about 120,000 farmers signed contracts in 1934 and 96,000 in 1935. Leaders in the Farmers Union had greeted with enthusiasm the decline in national enrollment in the second year of the AAA corn-hog plan.[12] In Illinois in 1934, approximately four million hogs were marketed by AAA signers and two million hogs were marketed by non-signers. In 1935 3.5 million hogs were marketed by signers and 2.6 by non-signers. The processing tax paid by non-signers was estimated to be approximately $12 million in 1934 and $13 million in 1935.[13]

Kennedy began to criticize the New Deal farm program in his writing and speeches. In June 1934 he told listeners to the *Farmers Union Hour* over the NBC Chain, "There are, in fact, just two programs.... One is the program of the National Farmers Union — the other is the program of the exploiters of the plain people." He said that the Farmers Union favored a policy of cost of production but that the agricultural program of the government did not recognize cost of production as the basis of price. "The present Agricultural Adjustment program constitutes no fundamental change from the program that has in the past taken the farmers' products away from him at less than cost of production prices. The form has changed but the substance of the method has not changed. The farmer is still selling his products for less than it costs him to produce them."[14]

He also criticized the New Deal in articles in the *National Union Farmer*, a journal of which he was the editor and chief writer. For example, in May of 1935 an article told of high Agricultural Adjustment Administration payments to the Metropolitan Life Insurance Company. A subheading read: "Evicted Farmers Forgotten While 500 Checks paid to Insurance Company."[15]

Another article was titled: "F.D.R. Silent on Farm Problems in Fireside Chat." The writer stated: "Evidently the President feels that agriculture has been lifted completely out of the slough and is headed safely down the high road of prosperity under the guidance of Destroyer Wallace and his staff of pig-killers and plowers-under."[16]

Kennedy also appeared with Milo Reno and Senator Huey Long at a meeting of the Farmers Holiday Association at the Fair Grounds in Des Moines, Iowa, on April 27, 1935. Huey Long said that the New Deal under Roosevelt was the "St. Vitus government." He commented, "What else could you call an administration that passes laws to get rid of the surplus when people are starving?"[17] Long found that his audience was very receptive. He told reporters later that day, "That was one of the easiest audiences I ever won over."[18] It is significant that both Long and Reno gave strong praise to Father Coughlin in their remarks at the Holiday meeting.[19]

At the same time that he was criticizing Secretary of Agriculture Henry A. Wallace and President Roosevelt, Kennedy was beginning to praise Father Coughlin and his National Union for Social Justice.

An article in the *National Farmers Union* told of a giant meeting in Detroit where the auditorium was filled to overflowing with 20,000 people, and tremendous crowds filled the streets outside the building where they listened to Coughlin and others on loudspeakers. Representative William Lemke was cheered when he complained about the "lunatic destruction antics of the A.A.A." Kennedy also spoke about the farm situation and received "thunderous applause."[20]

Kennedy had been working with William Lemke for some time writing and promoting legislation that they believed would be helpful to the farmers. According to his book *The Fed and the Farmer*, Kennedy had been one of the persons Lemke had turned to when the Frazier-Lemke Bankruptcy Act needed to be rewritten in order to stand up in the courts. Kennedy also worked with Lemke in attempting to achieve the passage of the Frazier-Lemke Refinancing Bill.[21] Writing in the *National Union Farmer*, Kennedy called the bill the most important part of the legislative program of the National Farmers Union. He said that he had worked to get farmers to put pressure on members of Congress to sign a petition to get the bill reported out of the Committee on Rules.[22] He gave a stirring speech over the National Broadcasting System on May 8, 1936, urging his listeners to write their congressmen to favor the passage of the Frazier-Lemke Refinancing Bill, without amendments.[23]

Kennedy had been influential in securing another potent source of support for Lemke. In the fight for passage of his refinance bill, Lemke received support from Father Coughlin. Coughlin had organized the National Union for Social Justice, at first with the idea of developing a pressure group to influence legislation.[24]

According to Kennedy, after the National Union for Social Justice was organized, Father Coughlin sent two representatives, Lou Ward and Fred Collins, to Washington. They were to find causes which they felt the National Union could support. Kennedy claims, "Now Mr. Ward and Fred Collins got ahold of me and ... I explained to them my program and I gave them copies of the bills, and Father Coughlin decided that that would become his program too. And I worked very closely with Father Coughlin. In other words, he worked very closely with me."[25]

Thus, through Kennedy, Coughlin became a supporter of Lemke and his legislation in Congress. Reeling from disappointment and anger after the defeat of his Refinancing Bill in the House, Lemke turned to an alliance with Coughlin. Lemke had not been able to defeat President Roosevelt in battles in the House of Representatives. Perhaps he could attack him through Father Coughlin and the National Union for Social Justice.[26]

Lemke had long been appealing to poor farmers with his various proposals, perhaps his appeal could extend into the cities. He received letters from supporters across the nation. One such letter came from C.E. Wheaton in Cincinnati, Ohio. He said, "I am one of the 11,000,000 Am. citizens not allowed to earn food and shelter while more than 3,000,000 Aliens are in the jobs that rightfully belong to Americans."[27]

A couple in Boone, Iowa, wrote, "Just a few lines asking you to stick to your fight for us farmers as we will go to the dogs if they isent something dun soon. We had our place clear of debts but since the new deal we are back under.... I am for the 16 principles what our radio preast preches to us. We are getting organized now, we will fight this through."[28]

In 1936 Father Coughlin changed the National Union for Social Justice into a political party, and in June asked Lemke to be its presidential candidate.[29]

E.E. Kennedy participated in Lemke's campaign in various ways. He helped Lemke get his name on the ballot in many of the western and middle western states.[30] Kennedy also gave many speeches for Lemke's candidacy. In a talk before the Union Party National Conference following the election of 1936, Kennedy said he had given 134 speeches in 26 or 27 states.[31]

Kennedy's support of Lemke was criticized by some of the members of the National Farmers Union. Several of the state Farmers Union constitutions contained provisions forbidding "sectarian-religious or partisan-political" discussions.[32] One group, the Washington-Idaho Farmers Union, passed a resolution stating that the Farmers Union had functioned successfully as a non-partisan organization for thirty-four years, and they feared that participating actively in politics would disrupt the work of the organization. They were concerned that the statements made by Kennedy as National Secretary of the Farmers Union would be taken as an endorsement of the entire Farmers Union organization for William Lemke as presidential nominee of the Union party. "While we may agree with Mr. Lemke's Platform, and no doubt a great many of our members will support him, yet our organization is an economic organization and not a political organization, and our program should be emphasized regardless of politics."[33]

In November of 1936 Kennedy was ousted from his position of National Secretary of the Farmers Union. This was because, his opponents said, he had actively engaged in promoting a political party, Coughlin's Union Party.

Later that year Lemke professed disappointment that he had received less than a million votes.[34] Lemke re-evaluated his plans, while Kennedy was licking his own wounds because he had lost the position that he had held as Secretary of the Farmers Union. The two men wrote several letters back and forth. (Kennedy always believed that he lost his position in the Farmers Union because of New Deal employees and supporters in the Union, and his resentment shows in these letters.) The two men agreed to meet in Chicago to discuss proposals for future action.[35]

Apparently at the suggestion of Lemke, Kennedy decided to move to Washington, where he established a legislative service supported by individual farmers subscribing for his services on the basis of $10 a year. He also published *Kennedy's Washington Letter*. In his first issue Kennedy roundly criticized the President's Court Packing Plan, and stated that recent court decisions on the Frazier-Lemke Farm Mortgage Moratorium Act showed the Supreme Court was the friend of the farmer.[36]

Following the Supreme Court decision in 1936 that declared the processing tax unconstitutional, farmers in Iowa organized to attempt to get the tax back and formed the Farmers Process Tax Recovery Association.

Kennedy read about their organization and their attempts to recover the process tax. He wrote A.J. Johnson, the secretary of the Association, suggesting that they hire him as their legislative consultant and lobbyist.[37] Johnson met with the president of the organization, Donald Van Vleet, and together the two agreed that Kennedy's help would be valuable in obtaining the return of the processing tax. A month or two later a group of the Iowa Farmers Process Tax Recovery Association leaders went to Washington to confer with Kennedy. Kennedy introduced the Iowans to a number of congressmen from farm states. The congressmen suggested that William Lemke and Kennedy work on drafting legislation calling for the return of the processing tax to the farmers.

Then in September 1937 the Recovery Association, on Kennedy's recommendation, decided to expand into other states, including Illinois. Kennedy had Farmers Union and Farm Holiday connections in Illinois that he could draw upon for selecting leadership in the Illinois Association.

The first person in Illinois to be chosen was Fred Wolf of Papineau, Illinois. Wolf had been president of the Illinois Farmers Union. In September 1937 he agreed to organize and supervise an association to recover the hog processing tax. For his efforts he was to receive five percent of the Illinois membership fees. If the processing tax was refunded he would also receive four-tenths of one percent of refunds to Illinois claimants.[38]

Following their original efforts, the state manager and most of the county recruiters experienced some setbacks. The organization did not grow as much as they expected it to. (The total membership in Illinois was approximately 900.)[39] Fred Wolf received a letter from Van Vleet in April 1938 stating that they did not have enough money to pay all their bills because of the added expense of the national organization. "Now Fred, I know you need this money but you can see just what position we are in here."[40] When Wolf wrote to A.J. Johnson, secretary of the Association, in May, 1938, he said that he had been spending most of the last month on his personal business.

How much time and money did the recruiters put into their work? The recruiters generally put in a day or two a week during most of the winter, and after planting returned to it for a few weeks during the summer before stopping work for fall harvest. They also had to pay their own gas and automobile expenses. Sometimes they paid for printed materials or took out advertisements in local papers.

One recruiter commented, "It is awful hard to get publicity in our Co. paper."[41] To this letter Van Vleet replied: "Keep after the county papers to give you publicity. This is news and it should go in…. Our program is going to gain momentum as time goes on."[42]

Other recruiters reported that newspapers of the area had published an article by Martin White stating that farmers could not recover their hog processing tax under the existing law. One recruiter, who had especially felt the effect of the Department of Agriculture articles, reported that some farmers from his area had asked him to return the checks they had given him for membership in the organization after they had seen the articles.[43]

A.J. Johnson tried to encourage his recruiters. He replied to a letter from recruiter Alfred Harm: "Sorry that you have so much trouble in your community with the Bureaucrats. I think the time is coming when you will have the laugh on them."[44]

In another letter, to John P. Lingenfelter, Johnson said the situation that Lingenfelter was facing in his community was quite general. The central office was receiving only a few scattered claims from different sections of the country. Johnson had heard from Kennedy, however, that the educational work the group had done regarding the process tax had at least made it "impossible for the Secretary of Agriculture Wallace to have the present Triple A financed by a processing tax. This being the case, I think we should feel, even though we have not as yet received a refund of the processing tax, that the money that has been spent has been a good investment."[45]

Shortage of funds was a problem that the Association struggled with during most of its existence. In the summer of 1938 Kennedy wrote to A.J. Johnson: "You will have to use your best judgment as to how the money is divided until we get out of the straight jacket we are or were in when I was there. On the other hand, if there is some that you can send my way, be it little or much, please do so."[46]

Some of the leaders in the Illinois branch of the National Farmers Process Tax Recovery Association were so excited about their work in the organization and so attracted by Edward Kennedy's speeches that they followed his lead when he suggested that they pull out of the Farmers Union and form a new farm organization.

Kennedy was angry at his expulsion from the office of National Secretary of the Farmers Union. He also protested the efforts of the National Farmers Union board to establish another Farmers Union organization in Minnesota. He said the board was violating Minnesota's rights.[47]

In May 1938 Fred Winterroth, Secretary-Treasurer of the Illinois Farmers Union and a solicitor for the Illinois branch of the National Farmers Process Tax Recovery Association, sent out letters to the State Presidents of the Farmers Unions of Michigan, Indiana, Illinois, Iowa and Minnesota. Members of the Board of Directors from Pennsylvania and Ohio received letters too, also E.H. Everson and E.E. Kennedy. These letters called for the leaders to meet in Kankakee, Illinois, on June 24 and 25.[48] In a later letter, Winterroth stated, "The consensus of opinion of all with whom this subject has been discussed is that definite action must be taken to preserve the principles for which we are organized."[49]

The meeting was held and a letter was written to John Vesecky, President of the Farmers Union. The Kankakee dissidents complained that the Farmers Union officials had not tried to make a determination of facts in either the Minnesota or Michigan case.[50]

A.J. Johnson of Iowa declined to attend the meeting. He did sign the letter to John Vesecky, at that time President of the National Farmers Union, protesting the suspension orders that had been issued to Minnesota and Michigan. The letter was signed by Robert Spencer of Indiana, J.C. Erp of Minnesota, Ira Wilmoth of Michigan, Fred Winterroth and Fred Huls of Illinois.[51] Then in August Johnson announced that he was not going to run for president of the Iowa Farmers Union. Kennedy wrote to Johnson that he was sorry to hear that Johnson was not going to be president for the coming year, because "there is so much to do and so few to do it. And the most important time in the Union is at hand in the next few months."[52]

Because of the controversy surrounding John Erp, President of the Minnesota Farmers Union and the National Farmers Union, and Erp's refusal to meet with the board, the national board of the Farmers Union withdrew the Minnesota Farmers Union charter. A few months later the charter of the Michigan Farmers Union was also withdrawn. The national convention of the Farmers Union confirmed these actions of the board in November 1938.[53]

Farmers Union members in a number of states were angry at the expulsion of the Michigan and Minnesota unions. The State Board of South Dakota heard objections from members who refused to pay their dues until the matter was cleared up. The Board directed that their complaints be published in the South Dakota Farmers Union paper.[54]

Also, there was objection to the National Farmers Union officers supporting bills for agricultural programs other than the cost of production plan. A resolution was passed at the District One Convention meeting at Vermillion, South Dakota, that the officers were "grossly misrepresenting the membership" in permitting the broadcasting of information to the effect that the National Farmers Union wanted the Pope-McGill Bill enacted.[55]

In January of 1939 Robert Spencer sent out a letter calling for a meeting to form a new farm organization. He wrote that it was important to get a new organization going to "help Kennedy and his coworkers in Washington get the Cost of Production bill, the Frazier-Lemke Refinancing bill, and the Hog Processing Tax Recovery bill made into law this session of Congress." The meeting was to be held at the Myer Hotel in Peoria, Illinois, the day before the state convention of the United Farmers of Illinois.[56]

Following these actions, Illinois, Indiana and Ohio pulled out of the National Farmers Union and joined Minnesota and Michigan to form the National Farmers Guild. The leaders in this Guild were often men who were also active in the National Farmers Process Tax Recovery Association. A letterhead of the National Farmers Guild in 1941 listed William E. Tanner as Secretary Treasurer of the organization, while the directors included Fred Wolf, Robert Spencer and Walter Meyne, all of whom had been solicitors for the Association. The letterhead also lists Edward E. Kennedy as the organization's Legislative Representative.[57] The group continued to support the principles of the old radical wing of the Farmers Union, including the hog processing tax refund.[58]

Kennedy had been the main motivator of this withdrawal from the National Farmers Union.[59] "He hoped that his group would grow," wrote historian Lowell K. Dyson in *Red Harvest*, "just as he and the other leaders of the National Farmers Process Tax Recovery Association had hoped that their group would grow. But they remained small. They were not strong enough to achieve the legislation for which they had labored. Their leaving had also weakened the Farmers Union which remained the smallest of the national farmers organizations."[60]

Kennedy continued as a powerful influence within the Farmers Guild for several years. Just as Kennedy's connection with William Lemke was evidenced in their collaboration in attempts to pass legislation authorizing a return of the hog processing tax to the farmers, so also his connections with Father Coughlin and the League for Social Justice were important in the Farmers Guild.

Father Coughlin, the Detroit radio priest, had long sought to appeal to the American farmer. In early meetings of the Union for Social Justice he had featured Edward Kennedy and William Lemke as "friends of the farmer." In his magazine *Social Justice* he carried many articles that would appeal to the farmer. Articles such as "New Tax Misleading to Farmers,"[61] "Who Signed the Frazier-Lemke Petition,"[62] "How Mr. Hull Stabs the Farmers,"[63] "Cost of Production: The Key to Lasting Solution of Farm Problem,"[64] "Farmers Tired of New Deal Quack Measures,"[65] "Wallace Report Sees Continued Farm Control,"[66] "Farmers Sour on New Deal Control of Crops"[67] and "New Deal Policies Bring Income Drop to Farmers"[68] appeared regularly in his paper.

Articles by Edward Kennedy appeared occasionally in *Social Justice*. One of these was an article on the Frazier-Lemke Bill, in which Kennedy quoted E.H. Everson, President of the Farmers Union, as saying that the Democrats could reuse their platform of 1932, since, "Except for repealing prohibition, they haven't used a bit of it." Kennedy stated that the Farmers Union would support candidates who believed in representative government and supported it. He felt, "The man who ought to be elected president is the man who will support the Frazier-Lemke refinancing bill and the 'cost of production' bill; who favors payment of the soldier's bonus with new money...."[69]

A particularly inflammatory article, entitled "The Farmer Goose-Steps," by Joseph P. Wright appeared in the March 7, 1938, issue of *Social Justice*. In the article Wright detailed some of the farm programs of the New Deal as "five years of experimental surgery in the New Deal clinic," which, despite five years of wasteful spending and crop destruction, had not solved the problems of the farmer. Then Wright warned that the American farmer had something even worse to fear: A New Deal law had been passed that gave to the Secretary of Agriculture "unlimited domination of five crops—wheat, corn, tobacco, cotton, and rice—on a nation wide basis. Farmers [who produced] these crops must comply with the Government's dictates as to acreage, output and market quota, or submit to a cash fine, after the notorious act has been affirmed by a two-thirds referendum. Collectivism has arrived in America."[70]

Letters from farmers were often printed in *Social Justice*. One such letter was from S. Fred Cummings of Mason, Illinois. He wrote that American labor was not the group that suffered most from "New Deal mismanagement." Instead, it was the American farmer who suffered most from deliberate "bear raids" on farm commodity prices. Cummings believed that Wallace was deliberately trying to destroy any possibility of farm recovery because Wallace wanted to establish an agricultural dictatorship.[71]

Articles in *Social Justice* praised farm organizations that opposed the New Deal. In the August 29, 1938, issue of *Social Justice*, in an article entitled "Farm Revolt Wins," the author stated that Secretary of Agriculture Wallace had announced that no steps would be taken to set up marketing quotas on the 1938 corn crop. "The decision is a victory for the Corn Belt Liberty League," wrote the article's author, "headed by Tilden Burg, of Sciota, Illinois, who led a farmers' revolt against plans of the A.A.A. to padlock Midwest corn cribs, the padlock being the 1938 version of prosperity by destruction."[72]

The Farmers Guild was also praised by *Social Justice* writers and editors. In an article entitled "These Farmers Have the Right Slant on Things," the editor commented on the resolutions adopted at the Michigan Farmers Guild Convention in Greenville, Michigan, in the fall of 1939. They had demanded the outlawing of the Communist Party, opposed a presidential third term, disapproved the Administration's neutrality, and the Administrator's agricultural and reciprocal trade programs. The writer commented: "The Farmers Guild deserves the plaudits of all true Americans for its unqualified opposition to Communism. In demanding that the Red party be declared illegal and removed from the ballot in the United States, the Guild recognizes the danger of protecting those who seek to overthrow the Government.... The sooner American citizens, particularly those entrusted with the reins of government, awaken to the perils of Communism, the sooner America will be assured of maintaining freedom; and the first step toward crushing the threat to American democracy should be elimination of the Communist Party from the ballot. Otherwise we are inviting destruction."[73]

Farm organizations in other states also experienced conflict concerning communism. At a meeting of the Ohio Farmers Union, Father Kaufman, one of Coughlin's representatives, criticized John Marshall, a leader of the left-wing Ohio Farmers League, who had hoped to win support for a proposed farmer-union party. Kaufman told the Ohio Farmers Union convention: "What we need is a people's party." He said that Marshall, the leader of the Ohio Farmers League, was an agent of international bankers and the Communist party, and that he had come to the meeting for the purpose of disrupting the Ohio Farmers Union. At this point Ohio Farmers Union men threw Marshall out of the meeting.[74]

One Farmers Union leader complained to Edward E. Kennedy that some of their Farmers Union members were spending all their time organizing for Coughlin's National Union for Social Justice instead of working for the Farmers Union. Several of their local groups had lost members because of this. He concluded, "If you are close to Father, wouldn't it be a good idea to have him urge his farmer members in some way [to] stay with the Farmers Union?"[75]

Related to Father Coughlin's fear of communism was a belief that communism and Judaism were connected. In an article in *Social Justice*, entitled "America's Insidious Foes," George Edward Sullivan began, "The time has come to cease evasion, shadow-boxing and deception about Communism, and to inaugurate a real investigation to ferret out and expose the precise identity of the occult forces behind it...." He then went on to discuss communism, stating that Nazism came into power in Germany in 1933 as an "antidote for Communism."[76]

The next foe that Sullivan discussed was nationalism formed along racial lines. He declared that Jews have divided loyalties and so cannot be wholly loyal to the country they live in.[77] The close proximity of the two foes in the same article makes it possible for a reader to conclude that they are closely related. Sullivan did not discuss Nazism or fascism as an enemy of America.

In another article in *Social Justice* the writer quotes from Hillaire Belloc's book *The Jews*: "The tremendous explosion which we call Bolshevism (now known as Communism) brought the discussion of the Jewish problem to a head. This movement was a Jewish movement, but not a movement of the Jewish race as a whole. The leadership was Jewish, but its current was not created by the Jew."[78]

In many of the articles in *Social Justice* the writers seemed sympathetic to the rising fascist movements in Europe. In the June 12, 1939, issue of *Social Justice* there is an article praising Antonio Salazar, the Prime Minister of Portugal: "Dr. Salazar is widely and justly acclaimed as a worker of financial miracles, as the statesman who has given his country a surplus in these years of almost universal deficits elsewhere, and who has founded the new Portuguese State, as a firm bulwark against Bolshevism, upon principles of Christian social justice."[79]

In another article, Mussolini is warmly praised. The writer states that Mussolini had never been proved wrong. "He made peace with the Pope. He saved the civilization of Western Europe in Spain. He has given his people dignity as well as the beginnings of an empire. He has yet to give them wealth."[80] In another article, the same author praises the actions of the Italian Fascist regime because it enacted a code that recognized Catholic marriage as the foundation of civil society, and also passed legislation which harmonized the school with the church by making the teaching of the Catholic religion the "foundation and coping-stone of education."[81]

The editors also refused to believe that the German government was persecuting the Jews, taking as their proof a speech by Hitler in 1934 in which he said that he had been protecting the Jews. Hitler had said, "But for me, there would not be a single Jew alive in Germany today."[82]

Probably the most blatantly anti–Semitic articles in *Social Justice* were a series of articles discussing *The Protocols of Zion*. These articles claimed that *The Protocols of Zion* were agreements by Jewish leaders to work toward Jewish domination of the world. Coughlin introduced the first chapter of *The Protocols* by stating that they corresponded with "very definite happenings which are occurring in our midst." He believed *The Protocols* were "preeminently a Communistic program to destroy Christian civilization."[83]

One *Social Justice* staff writer wrote, "The question is: 'Do the Protocols exist?' And the answer is: 'Yes.' The second question is: 'Do Jews in great numbers uphold Communism which definitely plans world revolution?' And the answer is: 'Yes.'" Then the writer suggested that if Jews did not want to be persecuted they should sever all identification with Communism.[84]

One Farm Holiday leader in Nebraska much later recalled that one of the women in their group had been a great supporter of Father Coughlin and often quoted from the articles on *The Protocols of Zion*. Harry Lux said, "She was sure that it was true in every way and that … Father Coughlin did the same."[85] And because of her prejudice against Jews she told Harry Lux in front of Tony Rosenberg that he should "steer clear of it [Jewish people]." After this, Rosenberg "just didn't go to any more meetings." Lux was certain this was because of the Coughlinite influence.[86]

Another example of anti–Jewish feeling came in the form of an outburst by F.C. Crocker at the 1935 Farm Holiday Convention. During a business session of the group, a representative of the Sioux Falls group read their resolutions. Reno interrupted the presentation and called on Crocker, who made an attack on "Communistic Jews." The Sioux Falls resolutions were then defeated.[87]

Several writers have suggested that Milo Reno, the leader of the Farmers Union, was anti–Jewish.[88] It does not appear that Kennedy was anti–Jewish, but he led farmers out of the Farmers Union and then allied himself and his group with Father Coughlin, who was. Kennedy left the Farmers Guild in 1942 when he accepted a position as recruiter and organizer for the United Mine Workers and their affiliate, the United Dairy Farmers.

After Kennedy left, Farmers Guild members turned to Carl Mote to lead their group. Carl Mote had written in *The New Deal Goose Step* about prominent Jews in Roosevelt's cabinet. Under his leadership the Farmers Guild would later become more anti–Jewish, but the seeds had already been sown.[89]

9

The Corn Belt Liberty League

In 1938 a new group arose, the Corn Belt Liberty League. It started when several farmers who lived near the town of Macomb, Illinois, talking to each other "on the street corner" after they received their corn acreage allotments, decided to call a meeting to protest the 1938 Agricultural Adjustment Act. Under the Act the government urged farmers to limit their corn production so as to help control farm prices. Each farmer was notified of his corn acreage allotment and asked to comply. A provision of the Act even allowed for marketing quotas with penalty taxes on sales above the established quotas if two-thirds of the farmers voted to impose them.

Farmers were already upset because farm prices had dropped nearly 30 percent below prices of a year ago, while the prices they paid for the goods they bought were only six percent lower. Now the government was asking them to lower the number of acres they planted in corn or wheat, which would work a financial hardship on many of the farmers.[1]

The Macomb County farmers called a meeting, expecting perhaps fifty or a hundred persons would come. But at the first meeting, held April 18, 1938, at the Macomb County courtroom, approximately 1,500 farmers showed up. They jammed the courtroom to capacity and overflowed into the corridors and out into the street. Those who were able to obtain ballots, distributed at the door, voted 866 no and 29 yes to the question: "Are you in favor of crop control?"[2] They adopted resolutions condemning the 1938 corn allotments as "absolutely ruinous to the farmer," and said that they would resist any attempt to force them to comply with the AAA regulations. They also elected officers; Tilden Burg, a livestock farmer from Sciota who raised corn to feed his cattle and hogs, was elected president of the group.[3]

A notice was put in the local papers: "Farmers! Urge your friends and

Farmers in town on a Saturday night. (Courtesy of Library of Congress, Washington, D.C.)

acquaintances from all parts of western Illinois to attend the district meeting of the CORN BELT LIBERTY LEAGUE, Wednesday, April 27, 7:30 p.m. at Macomb Armory. John E. Waters of Madison, Wis., a teacher of agriculture, employed by the Russian government during the original five-year plan and an authority upon farm regimentation, will speak. Also there will be special entertainment, music. If you want to retain the right to operate and control your own fields, if you oppose compulsory control from Washington, the Liberty League provides your opportunity to help in this struggle. Attend — Bring Your Farmer Friends."[4]

The leaders of the group drew up a list of objections to the government's new soil conservation and corn quota law. These were: The law denies the farmers the right to sow their land as they see fit, forces the breaking up of fields, ignores crop rotation, threatens a socialistic monopoly similar to Russia's, imposes special taxes which will ultimately fall on the farmers, favors the south, invites a surplus hay crop, and subjects American farmers to prices received in the world markets.[5]

Macomb, Illinois, is located not more than fifty miles from the Iowa border. Farmers from several corn belt states arrived in Macomb for the meeting.[6] Preparations were made to accommodate the expected crowd at the Armory. A public address system was installed to enable the audience to hear the speakers. The street outside the Armory was roped off so that farmers who could not find seats inside the Armory could stand within hearing distance.[7]

The meeting held at the Armory in Macomb attracted an attendance of 5,000 or more. Burg conducted the meeting. He told the crowd, "In recent weeks we have received corn allotments that for many of us are plainly ruinous. Everywhere we hear the charge, honestly made, that favoritism has influenced the fixing of allotments. We have seen compulsory control fastened upon farmers in other sections. We have the same sort of dictation awaiting us unless we organize and fight."[8]

He said, "The purpose of our organization is to oppose in every honorable way the un–American program of compulsory crop control which is being forced upon us.... The most of us have spent our entire lives on the farm. We have an affection for our lands that probably cannot be understood by politicians in Washington. Our farms will not be the same, when, instead of planning their operation along sane lines, we have jobholders coming from Washington to tell us what we can plant and what we cannot plant, perhaps under threat of prosecution and penalty."[9]

Rev. H.M. Bloomer, who also owned a farm in nearby Mercer County, was among the men who spoke to the crowd. He recalled that the President had said that one-third of the people of the United States were not fed, clothed or housed properly. That being the case, he demanded to know why production was being curtailed.

Rev. Bloomer said that when he was in Mercer County a few days earlier he saw several hundred bushels of corn waiting shipment by barge to New Orleans. He was told that the buyer had learned that corn prices in New Orleans had recently fallen because of great shipments of corn coming in from Argentina. The buyer could not sell U.S. corn in New Orleans at a price to compete with Argentine corn and still make a profit. Bloomer concluded, "What we need is trade relationships that will aid our farmers." This remark was greeted with loud applause from the audience. But someone with a different interpretation of the reason for the piling up of corn in Mercer County spoke up. At the conclusion of Rev. Bloomer's remarks he called out that the reason the corn was awaiting shipment was not because of the price of corn in New Orleans but because of labor troubles.[10]

John E. Waters of Madison, Wisconsin, who had been a power farming instructor in Russia for a number of years, told the group that the United States agricultural program was becoming more like the system existing in Russia. He said, "You will find yourself gradually drawn into the web of Communistic regimentation with loss of life and liberty if you permit the government to continue its dictatorship of your farm operations." He also charged that the crop control act would create spies among the farmers and breed hatred between farmers and their neighbors. "This

New Deal," he charged, "is snipping off a little bit of our freedom every day. Unless this tendency is checked our government will become as bad as Russia's."[11]

R. Lowell McDaniel of Greenfield, Indiana, who was president of the Indiana Federal Conservation Club and secretary of the Hoosier Taxpayers Union, said that he had come to the meeting because many of his farm friends had asked him to find out about the Macomb movement and bring the story back to them. He believed that ninety percent of the farmers in Indiana were against the government farm bill.

Burg charged that the farmers had not been consulted in forming the latest legislation. "We have not been truly represented in Washington. Farm organizations have spokesmen and lobbyists in the national capital, but do these spokesmen express the views of yourself and your neighbors? They do not. Perhaps they represent the views of the minority of farmers, but we are convinced beyond any doubt that the majority of farmers do not want this sort of thing."[12]

There was an opportunity before and after the meeting for local farmers to pay $2.00 dues and join the Macomb Corn Belt Liberty League. It was announced at the meeting that over 350 farmers had joined the group.

Burg said that he was surprised at the overwhelming response the group had received. A roll call from the platform that night showed delegates were present from Indiana, Ohio, Iowa, Missouri, Minnesota, Wisconsin and a number of counties in Illinois. Emphasizing the importance of numbers, Burg told the visitors to "go back home and organize to fight this thing. We'll help you in every way we can."[13]

The protest meetings led to a spirited discussion in Macomb on the government's agricultural program. The *Daily Journal* printed several letters to the editors from local residents: Mrs. Frank Taylor wrote she believed that the majority of the supporters of the movement were "such strong Republicans they hate to accept anything, good or bad, proposed by the Democrats."[14]

Mrs. Tom Joy replied that she was also a farmers wife and was fully aware of the struggles of a farmer, "as we have been married only eight years, during that time there has been depression, drought, recession, etc.... I wouldn't say that strong Republicans were the only ones who attended the meeting Monday evening, if so the dear old party is surely slipping in McDonough county. Yes, Mrs. Taylor, I think most of those people had forgotten about their party." She continued: "My husband farms more land than Mr. Taylor but his corn quota is less, but then one must consider the fact that my brothers-in-law aren't members of the board of directors of the Farm Bureau.... I would think crop control was simply grand, Mrs. Taylor, if we had the same arrangement as you folks. But what would you think of it if you had about sixty acres that would have to be put in beans because the grasshoppers killed the clover last fall and you were allowed to plant sixty-three acres of corn when you were planning on one hundred."[15]

Another farm woman, Mrs. W.L. Heberer, wrote that she wanted to know who counted the votes, "Will there be an equal number of 'unbelievers' to count the votes (in a corn referendum) or will the county committee (who to a man, are for the New Deal, on account of a good job) do the counting?"[16]

The excitement stirred up by the meetings in Macomb reached other communities in Illinois. Tilden Burg announced that several of the leaders of the Corn Belt Liberty League planned to go to Monmouth to help

Feeding chickens in Jasper County, Iowa. An Arthur Rothstein photograph. (Courtesy of Library of Congress, Washington, D.C.)

with the organization of the Warren County Corn Belt Liberty League there, and they were also planning to go to Princeton on May 7. He said farmers of these areas had asked for help in organizing local groups.[17]

Five hundred farmers attended the first meeting of the Warren County Liberty League. They also adopted resolutions similar to those drafted at the Macomb meeting. These resolutions demanding repeal of the crop control measures were sent to President Roosevelt and Secretary of Agriculture Henry A. Wallace.[18]

At Princeton 500 farmers met in the Princeton High School auditorium and voted unanimously to oppose the government's crop limitation program. Tilden Burg, president of the Macomb Corn Belt Liberty League, attended the meeting and announced that he had received "hundreds of letters from farmers all over the country ... and they are urging that they be given a chance to protest in some effective way against governmental control of the farmers' business."

Another speaker was Arthur H. Booth of Princeton. He blamed the Farm Bureau for its promotion of the unpopular law. He said, "I blame the bureau for having this law passed. I don't think the measure suits the farmers—certainly not the corn farmers. It may be all right for cotton producers, it's all wrong for us."[19] Booth used the example of a cooperating farmer who would get only $663 from 1000 bushels of corn, including his benefit payments, while the non-cooperating farmer would get $717 for his corn. Booth also referred to the cost of production bill, which he called "a real farm bill." It had been defeated by only a few votes, he said, in both the house and the senate.[20]

Carl Dawson, who was elected president of the local group, asserted, "This law is being jammed down our throats and we are going to oppose it. It is a step toward dictatorship."[21]

All this response was a bit overwhelming for the farmer leaders of the Macomb Corn Belt Liberty League. Burg told a reporter that he was getting more letters with every mail. There were letters from farmers in Iowa, Missouri, Indiana, Ohio, Illinois, Minnesota and Wisconsin. Nearly all endorsed the movement. Some of the writers asked for literature and others wanted to know if speakers were available for organizational meetings.[22] Stanley Morse, a leader of the American Liberty League and the Farmers Independence Council, wrote to Burg. He congratulated him on the "splendid work" that they were doing. He also suggested that they keep the control of the organization in their "own hands and [not] get mixed up with political parties."[23]

They decided to open a headquarters for the Corn Belt Liberty League. They rented an office in an upstairs room in the Illinois Theatre building in Macomb. They also hired a part-time secretary. An announcement of the opening of the headquarters was printed in local papers: "Persons interested in information regarding this organization and those who desire to lend their support to this movement which is designed to preserve the independence of the American farmer are invited to visit us."[24]

Since farmers usually have much to do in late spring and early summer, Burg told reporters that he and his fellow officers were finding it necessary to work extra hours and perhaps let farm work slide a little in order to carry on the program of the Corn Belt Liberty League.[25]

Other leaders also spent considerable time traveling around to the various meetings getting groups started in other counties. Mrs. Carl Ruebush remembers that she and her husband often left early to travel together with other farmers and their wives to attend meetings. She told her interviewer that they went to all the meetings. "And when I think back the way he, he was such a good farmer, and everything. I think back and I just wonder at him leaving as early as we did some nights to go so far off ... because he would have so much to do at home."[26]

Mrs. Ruebush remembers that she and her husband, Carl Ruebush, and Finley Foster and his wife, traveled to many meetings, including a meeting at Princeton, Illinois, 120 miles away.[27]

The Foster and the Ruebush families had been friends for years. Foster was fiercely independent. He even put up a sign in front of his place: "THIS FARMER IS NOT ON GOVERNMENT RELIEF." When asked about it he would reply, "I can make a living without government help.... I'd rather run the farm myself. I hope they won't take it away from me.... I'd rather sell 200 bushels of corn at 40 cents than 150 bushels at 50 cents. I'd still be $5 to the good. The farmer ought to know how to keep up his farm without government help. This program is penalizing the rest of us to help the farmers who have worn out their land. Every man would get along if he'd spend less than he takes in."[28]

The trips that the Fosters and the Ruebushs took for the Corn Belt Liberty League were at their own expense. Carl Ruebush was a director for the League, and he did not get paid for that either.[29]

The group continued to spread. In late June Burg announced that meetings would be held at Paw Paw, Aledo and Lewiston. A meeting was also scheduled to be held in South Dakota at Mitchell. He announced that the League headquarters had received notice that 174 members had joined the Edgar County Corn Belt Liberty League. Second meetings were scheduled for Adams County and Marshall County in Illinois.[30]

Attorney Frank Dick spoke at a meeting at Mendon, Illinois, telling the farmers, "The control of your farm has been moved from here to

Washington, D.C., given over to the exploitation of some favored foreign nations. I never did believe that we had 'overproduction' in the United States and I now have plenty of government statistics to support my contentions—and I challenge anybody to refute them."[31]

Tilden Burg said that he feared for the future. "They get their foot in the door with the compulsory feature of this act, and next year it will be worse. Then the way is open for complete control over our farms by officials. That's what we want to head off right now."[32]

At a June 10 meeting at Toulon, Illinois, the farmers adopted a series of resolutions: "Whereas, the farmers of Stark county and the surrounding territory ... look with fear and disfavor on certain features of the crop control law and its administration by the United States Department of Agriculture, and ... are convinced that the present agricultural law is so designed that it can be used to bring about, through threats or inducements, the adoption of compulsory features which strike at the very heart of individual liberty and the farmer's right to operate his farm as he sees fit ... [they] are determined to resist to the last, any efforts to force them, under the threat of prosecution, to give over the control of their lands to other people...."[33]

As the leaders of the Corn Belt Liberty League continued to fight the New Deal agricultural program, they began to think through ways they could resist it. They admitted that the law as it was administered was not compulsory, but they said it contained a provision for compulsion.

According to the provisions of the law, a farmer choosing to exceed the acreage allotted to him by the AAA would not be eligible for benefit payments on reducing soil depleting crops, and could get only 60 percent as much corn loans as a farmer who complied with his allotment.

Furthermore, if corn growers in the Middle West exceeded national production goals, penalty taxes could be applied if two-thirds or more of the farmers affected voted in referendum to apply marketing quotas. If adopted, this would impose a penalty tax of 15 cents a bushel on corn marketing in excess of the quota assigned to each farm.

Burg said that the members of the Corn Belt Liberty League were "opposed to such compulsion and we believe that a majority of Corn Belt farmers are opposed to it and will vote against it if they are fully informed and will take the trouble to vote. One of the purposes of the league is to tell farmers what a marketing quota will mean and to urge them to vote in case a referendum is ordered."

He further commented: "In the educational meetings that the AAA officials have conducted in this section those of us opposing the corn program have had no effective voice. We could attend as individuals and express our opinions, but they had no effect against the organized propaganda of the professional AAA officials and the paid county and township committeemen. We intend to build an organization strong enough to make ourselves heard."[34]

In July the group decided to publish a journal, the *American Liberty Magazine*. It was suggested by Clifford B. McGrew, an insurance agent from Galesburg, Illinois. The Corn Belt Liberty League board of directors voted to establish it, and the first edition was published on August 4, 1938.[35] League members received three complementary copies, and then it was hoped they would pay to receive additional copies. The price for non-members was one dollar a year, for members it was seventy-five cents. The magazine was full of articles about Liberty League activities, and letters and articles supporting their stand on agricultural issues.

The movement began to expand outside of Illinois. One of the next states to be affected was Indiana. A few organizers from Indiana had attended the second meeting of the Corn Belt Liberty League held at the armory in Macomb. Lowell McDaniell of Greenfield, Indiana, who had attended the Macomb meeting, was one of the early leaders in organizing the Indiana Corn Belt Liberty League.

Republicans also began to organize a similar group called "Republican Farmers of Indiana." Spokesmen declared that it was designed to offer a forum for corn growers to formulate a united protest.[36]

In July a large meeting of the Corn Belt Liberty League was held in Indianapolis, Indiana, at the Claypool Hotel. The group heard addresses from Colonel Walker Garrison, president of the Associated Farmers of the Pacific Coast; Wallace Walker, district attorney, Macomb, Illinois; Stanley Morse, organizer for the Farmers Independence Council; Rev. H.M. Bloomer of Macomb, Illinois; and Lowell McDaniell of Indiana, who called the meeting.

McDaniell announced that the organization of the Indiana Corn Belt Liberty League was moving swiftly. They had set up state headquarters in the Claypool Hotel and were sending out organizers to form district and county organizations. A great number of memberships were coming into the office daily, many of them unsolicited, showing the growing interest of farmers in the organization.[37]

When G.C. James, secretary of the Corn Belt Liberty League, Rev. Henry Bloomer and Wallace A. Walker returned home to Macomb from the meeting in Indiana, they reported they had enjoyed a good meeting and that a major concern of the group seemed to be for an organization of farmers on a national scale to combat activities to control agriculture.[38]

In Nebraska a group called the Grain Belt Liberty League was formed.

The president of the group was Carl Teft, a farmer who operated a 480 acre farm in Cass County. Teft had opposed the adoption of the compulsory marketing quota feature of the agriculture bill.

He told farmers that what the country needed was not less agricultural production but more. There were many people in the United States who were not well fed and clothed. The answer to their needs was not reduced production but an increased program to create jobs. Teft was also very interested in developing new uses for agricultural products. He felt that the government should support scientific research, such as that favored by the Farm Chemurgic Council of Omaha.[39]

The author of a later article in *The Nebraska Farmer* quoted AAA scouts as saying that the revolt in the corn belt and trouble in the cotton and tobacco belts have "subsided again." However, the writer of the article suggested that the unrest was continuing: "Farmers are waking up to the fact that they may be called upon to vote for or against compulsory storage quotas in corn."[40]

In Kansas Dan Casement, former president of the Farmers Independence Council, heard about the Corn Belt Liberty League from Morse. Morse wrote Casement about the group. He said that the group had been organized by "real dirt farmers." Morse was pleased to report that one of the organizers, G.C. James, had been a Farmers Independence Council member. Morse felt that the seeds planted by the Council had been part of the reason for the development and growth of the Corn Belt Liberty League.[41]

Soon Casement began to receive letters asking him to speak to local Kansas groups. Thale Skovgard, a member of the Kansas State Senate, wrote asking him to speak in Washington, Kansas, and to help organize a group "on the order of the Corn Belt Liberty League."[42] V.M. Reed asked him to stage a meeting in Ottawa County. He said he believed the feeling in his district was strongly against the AAA. "The vote against the proposed soil conservation district was 930–204."[43]

C.L. Potter of the Exchange National Bank of Clyde, Kansas, asked Casement if he would come to a meeting in Cloud County. He said that the farmers in the county were angry because the Department of Agriculture had mailed out a bulletin to the Farm Bureau office saying that all farmers should come in by June 1 to give information on their crop yield and acreage, since "all farmers in the county may at some future date operate under compulsory wheat quotas."[44]

H.E. Gordon of Horton, Kansas, wrote Casement that he was interested in opposing the farm program and had heard that there were meetings being called in other states among farmers who also were opposed. He asked Casement who they could get to come to Brown County to "expose the deception of the Bill and how it may be construed to mean anything its proponents desire."[45]

A friend wrote Casement some advice, which he seemed to follow as he related to these new developments. Ballard Dunn wrote advising that he stay clear of the Liberty League of the duPonts, because, "That organization is a mistake. The only organization that can make a dent is one that starts from the bottom without any 'big names.'"[46]

Casement also received a letter from George Herzog of the Kansas Taxpayer Association asking him to speak at a meeting of their group on May 28. Herzog wrote that Casement had been recommended by Senator Skovgard, with whom he had shared the podium at the Washington, Kansas, meeting.[47]

One of the leaders of the Washington, Kansas, group wrote to Tilden Burg, the president of the Illinois Corn Belt Liberty League at Macomb, Illinois. He wanted to know more about organizing a similar group in Kansas. Burg answered the letter and suggested that they get about 50 to 100 members before calling a larger meeting. "We have found that a set-up of this kind is very effective in encouraging membership ... what we need is a large number of members for the success of the organization."[48]

Ed Harvey of Lawrence, Kansas, wrote that they recently had held a meeting of farmers at which "Mr. Williams of the College explained the bill as Mr. Wallace told him to. Now what we want is for you to come down and tell us in our own language the cussedness of the bill and how it makes for the destruction of Agriculture in the U.S."[49]

Casement wrote Morse telling about the farmers' meetings that were springing up in Kansas. Morse replied that similar meetings were being held in Indiana, Ohio and Nebraska. He said that he had talked with the leaders of the Corn Belt Liberty League. "They are genuine farmers who are overwhelmed by the success of their move. If this move can be given planning and guidance, it is likely to sweep the Corn Belt."[50]

H.E. Gordon wrote Casement again and mentioned that one of the men in the Farm Agent's office told him that "they were not losing any sleep over the protest meetings. [Because] they were just a few bellyachers on a spree."[51]

Other persons who contacted Casement concerning speaking at Kansas farmers' meetings included J.J. Lilley of the Kansas Farmers' Holiday Association and George A. Hunt of Wellington, Kansas, Secretary of the Anti-Farm Control Association. Hunt said they had men carrying a protest petition to farmers in almost all the townships of the county.[52]

A.J. Ostlund, who had organized the first meeting at Washington,

Kansas, sent Casement a copy of the newspaper report of their organizational meeting and also the resolutions that the group had adopted. He said the resolutions were nearly the same as those sent by the Corn Belt Liberty League of Macomb, Illinois.

Ostlund also enclosed a letter from Charles Kishop, one of the organizers for the League, who had been setting up a meeting in Clifton, Kansas. Kishop wrote that he had arranged for the bandstand for the meeting and would have seats furnished by the lumber company. He wrote, "If weather is not fare [sic] will use the school house."[53]

Morse wrote Casement in late May that other units of the Corn Belt Liberty League were being organized. He commented: "The farmers seem to be determined not to comply [with the Farm Bill)]." Morse had been working with farm groups who were opposing efforts by labor unions to organize cannery workers or creamery workers. He had developed some ties with the Associated Farmers of California who were fighting unions also. He hoped that the movement would spread rapidly and "put labor in its place and deal the N.D. [New Deal] a vital blow." He commented, "the farmers certainly are getting fed up. We must work out a plan whereby the Corn Belt revolt is coordinated with the Assoc. Farmers move, any suggestions?"[54]

A few days later Morse wrote another letter to Casement. He wrote that he would like to come to Kansas to help Casement with organizing the Corn Belt Liberty League if enough financial backing could be arranged. Referring to an earlier letter of Casement's, he commented, "The lack of publicity you mention I am coming to believe may be a good thing. If the meetings continue to be held and organizations set up, the fire will be spreading quietly but steadily until it suddenly bursts out into a big conflagration and it then would be too late for the New Dealers to put it out or even stem it. There is a public feeling that the movement is dying down, although in Illinois and elsewhere meetings, are being constantly held and organizations set. If the movement can be kept going without too much noise for a month or two and all gains well consolidated, it should become very powerful."[55]

S.M. Swenson, Texas rancher and New York corporation president, who had been one of the supporters of the Farmers Independence Council, wrote Casement that he was encouraged to learn from Casement's letter that "many of the farmers are beginning to see how rotten and harmful to their own interests is the Farm Bill and to what this kind of control would eventually lead. Certainly there is no place where work such as you are doing is more needed and where, when successful, it will have the greatest effect politically than among the farmers."[56]

A meeting was held at Manhattan, Kansas, in early July to discuss the possibility of setting up a statewide organization to fight the 1938–1939 Agricultural Adjustment Act.[57]

Swenson wrote that Morse had written him that he needed money to go to Kansas to help with farm organization work there. Swenson sent Casement a check for $250 to use in any way he saw fit.[58]

Casement continued to receive invitations to speak at various meetings in Kansas. Helen Lobdell wrote to invite him to a picnic at Nickerson, Kansas, where he would be part of a program that would include the Republican candidates from that area. "We are most grateful that you will come to us for we feel your message is just what the farmers of Reno Co. as well as some others need." V.E. Becannon of Buffalo, Kansas, invited him to speak at a meeting at Fredonia. Casement was also invited to a picnic held by the Farmers Union in Riley County at which E.E. Kennedy would speak. E.E. Kennedy had once been a power in the National Farmers Union and was active in the movement to recover the hog processing tax. Now he had come to Kansas to protest the crop reduction program of the Agriculture Department.[59]

A state office and headquarters of the Kansas Farmers Liberty League was established at Horton, Kansas, with H.E. Gordon as president of the group. Petitions and membership cards were printed up by the state group.[60]

Morse wrote Casement a long letter stating that he was planning to come to Kansas to help with the campaign there. "My present plan is to leave here [Chicago] for Kansas City at 8.45 P.M. Wed. Have in mind going to Topeka in A.M. as I want to talk to Cogswell and one or two others and feel them out. Then I shall arrive in Manhattan on the streamliner about 7 P.M. ready for the fray."[61]

Morse enclosed a suggested letter to Senator Josiah Bailey of North Carolina, whom he thought would make a good speaker at a giant anti-farm program meeting to be held somewhere in Kansas.[62] In a letter written a day later, Morse wondered if he should go first to Washington to see if Bailey would be interested in coming to Kansas. If they couldn't get Bailey or some other Democrat who was opposed to the agricultural program, Morse suggested they might try to get Senator Bridges, "who was formerly a county agent and is a forceful speaker." Morse continued: "Plan is first to line up a speaker and then get busy on building up a rousing big meeting."

Casement replied: "Letters received. Think best you come here and consider local situation before contacting prospective speaker. I go to Cowley County Friday. Better accompany me."[63]

Morse traveled to Kansas and adjusted his methods to the local situation. He wrote former President Hoover that they were planning to hold a state convention representing 25 to 30 county units to set up the Kansas Farmers Liberty League. He commented, "The rapid growth of this movement is exceeding our expectations for we had not realized to what extent the farmers were waking up…. I wish you could get over here to talk with some of the splendid leaders who are emerging here."[64]

Casement's name was beginning to appear in Kansas newspapers again. In an article in the *Kansas City Star* entitled "Fire on 'Wheat Bribe,'" Casement was quoted as saying that "farmers will not cooperate in the minimum acreage wheat plan [because] they now clearly comprehend the sinister purpose of the Washington bureaucrats and will refuse to barter their liberty for bribes."[65]

In late August Secretary Wallace announced that no steps would be taken to set up marketing quotas for the 1938 corn crop. In Illinois Tilden Burg jubilantly announced the reversal of policy. "I think Wallace knew he would be unable to get the two-thirds majority vote in any referendum in the commercial corn area that would be necessary to make the marketing quota effective." He added, "League members and other farmers who planted corn to suit their own needs are going to be mighty proud of the extra crib of corn they will be able to harvest this Fall from the 'forbidden acres.' We will now have plenty to feed our livestock, and if there is any left over it will keep. It will come in handy next year if we should have a short crop, as we did in 1931 and 1936. We might have some corn to sell to unfortunate farmers who followed Washington orders and restricted their acres."[66]

Casement continued to receive letters from people asking him to speak at farm meetings. He was asked to be an officer of the Leonardville, Kansas, group. Victor Hawkinson of the Farmers Union asked him to attend a meeting at Council Grove. J.W. Wilson asked him to come to Arkansas City, Kansas, to speak on the farm bill. Conrad Crome asked him to come to help them organize a group at Marysville, Kansas.[67]

Morse wrote Hoover that he liked his Kansas City speech very much and felt that many others did too. His experience of the past few months had convinced him that ordinary citizens welcomed "militant, battling, truth-speaking leadership."[68]

Reports of the growing strength of the opposition to the AAA began to circulate in and beyond Kansas. The Kansas State Board of Agriculture invited Secretary Wallace to come to Hutchinson to discuss the federal government's wheat program.[69]

Dan Casement went to the meeting at Hutchinson, Kansas, in late

September. He stood up to speak at the meeting and told Wallace and the audience: "You cannot have a planned economy in a democracy and the AAA is a planned economy, which if carried to its logical conclusion, means the destruction of democracy. If the AAA benefits were discontinued, the AAA would blow up overnight. You are bribing the farmers to give up their liberty." Casement's remarks were greeted by a storm of booing from the audience. And when he asked those in the audience to stand who opposed the program, only two others rose to stand with him.[70]

Following the meeting he received many letters from farmers supporting his position. One farmer wrote: "I see by the paper that there were only three or four farmers out to the wheat parley. I believe the real farmers never take time to go to such places, and if they do they are easily lead by some silly emotional speaker, so they prefer to stay at home. Glad to hear you were there and was not surprised that the handpicked crowd booed you."[71]

Fred Grantham of Trenton, Missouri, wrote: "I agree most heartily with everything you have said. I note they called [unclear] you at Hutchinson, Ks., a Wallace meeting, but keep up your fight."[72]

W.F. Thompson wrote that he had heard over the radio that there had been an overwhelming rejection of Casement's plan to abandon the program. Thompson had noticed reports in the local paper that many of the people connected with the county agent's office or other offices of the farm agencies had been out of town at Hutchinson. "I'll hazard a guess that they pulled a fast one and packed the meeting at Government expense."[73]

G.W. Norris of Hutchinson, Kansas, wrote that he had been at the meeting but had left after Wallace had finished his speech. "I considered he had told but one side of the question and that no opertunity [sic] would be given to any one to present anything else."[74]

Elias Farr wrote: "I note in the *Kansas Farmer* that a roar of booing followed Mr. Casement's statements. Shucks that nothing. At those meetings they always have about 20 self-interested tax eaters to one real farmer. I was to the first meeting Wallace held in Salina. I think in '33, the last days of June. It Seemed as if everybody was there except farmers. Our good governor was their, officers of the state schools was there. All the big shots and little Cozack pots from the Agriculture College. Virtually all the county agents and their assistants of this state. County agents from north Oklahoma, from East Nebraska, from West Missouri was there. Yes I actually seen two farmers there. They had a breakdown on their combine and made a hurried trip to Salina for repairs and stopped in at the meeting for a short time. Yes Mr. Casement, the farmers were all Busy harvesting so they could pay the expenses of that motley horde of tax eaters. So you see they always have a large supply of Hooters and Booers."[75]

Casement's friend, Thomas Doran, wrote: "Had you used your wits at Hutchinson when you called for a vote, you should have limited the ballot to farmers who were not on the payroll of the government. You might as well go to Hell for a snowball as to undertake the county agents and managers of the new program to vote against it. It just cannot be done, and you need not be discouraged, but must glory in your strength and courage in facing the animal in his lair."[76]

C.S. Walker of Macksville, Kansas, wrote that "as in all New Deal meetings, they were unfair to their opposition. That is always their game, to down the opposition, by means fair or foul." Walker also wanted to know more about organizing a Farmers Liberty League in his area.[77]

Invitations to speak continued to come in. One group had arranged for Casement to speak at the sales pavilion in Arkansas City, Kansas. He also was invited to speak at a "farmer's non-political rally" at Independence, Missouri, held by the Independence Republican Club. Thale Skovgard invited him to Hanover, Kansas, for a meeting of the Kansas Farmers Liberty League of Hanover. "Come if you possibly can. It's the enemy's territory."[78]

Morse wrote that he had a very satisfactory talk with Hoover the day before. "He agrees with you and me that a militant, two-fisted leadership and fight is only way to save country." Morse also suggested that the Kansas Farmers Liberty League put all candidates for Congress on the spot and ask each of them in a letter if they would work for the repeal of the Farm Control Act.[79]

Casement continued to speak at various meetings up until the November elections. All these activities of the Corn Belt Liberty League, the Nebraska Wheat Belt Liberty League and the Kansas Farmers' Liberty League had not gone unobserved by the Department of Agriculture. The Department had responded in various ways.

Even before the Corn Belt Liberty League was formed, in January of 1938 the Department of Agriculture issued a statement that the farmers could not expect the gains in their cash farm income to continue. Secretary Wallace warned that surpluses of agricultural products were such that they threatened to engulf food and fiber producers. Wallace said that the farm production in 1937 was the largest on record. The Department of Agriculture forecast a smaller domestic demand for farm products (because of declines in industrial employment), lower farm prices (because of last year's bumper crops) and increases in real estate taxes.[80]

Writing to defend the 1938 Farm Bill, Senator Norris wrote to C.L. Dietz, Master of the Nebraska Grange, that "the loss of 'liberty' which many farmers regret will 'prevent a catastrophe.'" Quoting Norris' letter,

McKelvie, the publisher of *The Nebraska Journal*, wrote that this indicated the American form of government was about to be "scrapped." McKelvie believed that farmers were becoming aware of the seriousness of the threat that was being made to their independent action in managing their farms.[81]

When the first meeting was held at Macomb, one person tried to defend the legislation but was howled down by the crowd.[82] Following the meeting, the Farm Bureau, which had enthusiastically supported the various agricultural adjustment acts of Roosevelt's New Deal, branded the whole movement as political. A spokesman for the Farm Bureau said that only one of the leaders of the Liberty League was a farmer. And they said farmers had continued to sign up with the AAA even after the League had been organized.[83]

In Macomb the County Farm Bureau announced that they had received a statement from the Illinois Agricultural Association that corn might be selling for only thirty-five cents a bushel if farmers planted their usual acreage. But if they reduced their acreage to keep within their allotments, supplies would be held down to the point at which loans would be assured up to 75 percent of the parity price. "At present this would mean approximately 63 cents per bushel. Add to this the 10 cent per bushel payments for compliance, and corn growers would stand a chance to net better than 70 cents a bushel for their 1938 crop."

Thus, according to the writers of the statement, farmers had a choice: "to cooperate in making possible a return of 70 cents or more per bushel or to assist in piling up another huge surplus with disastrous prices such as we had in 1933."[84]

On May 2 Representative August H. Andresen addressed the House of Representatives. He spoke about the headlines carried in the newspapers of "Crop Rebellion, "Farmers Fight Compulsory Control," "Thousands of Farmers Join Corn Belt Revolt Against Government Crop Control," and the reports that told of spontaneous mass meetings attended by thousands of farmers in opposition to the crop control program as administered by the Secretary of Agriculture.[85]

Andresen said that this was not a revolt against the constitution but rather an "uprising on the part of thousands of honest and patriotic American farmers in defense of the constitutional freedom to live and operate as American citizens under guarantees set forth in the Bill of Rights."[86]

He said that thousands of letters, telegrams and telephone calls had been sent to members of Congress during the past ten days from farmers in the corn belt protesting the individual farm corn acreage allotments made by the Department of Agriculture. AAA regional administrator

Claude Wickard had reported that, "complaints against reductions in corn acreage allotments have been as high as 1,100 per county."

Andresen quoted from a Minnesota farmer who wrote: "Most of the farmers around here are opposed to cutting down their corn acreage. We ourselves are cut from 65 to 26 acres and others accordingly. The reduced acreage will not supply sufficient feed to take care of my hogs and dairy herd, and I will be forced to sell a part of my livestock this fall for lack of feed."

Andresen said that the Department of Agriculture was planning to begin an immediate drive to quash the revolt of the corn belt farmers. "It is also understood from good authority, that the Department dispatched several Secret Service operatives to attend the Macomb, Ill., meeting which was held by farmers from six states in the area."[87]

But the AAA representatives and Farm Bureau leaders said that the revolt originated in the towns rather than the farms, that it was politically motivated, only existed in small special localities, and had little or no influence. They said it had been accorded publicity far out of proportion to its real significance.[88]

The Agricultural Adjustment Administration also announced that their field representatives would hold "educational" meetings in an attempt to win back the support of dissenting corn belt farmers. Claude Wickard blamed "spell binders" and "misunderstandings" for the protests. Denying that the new crop control law constituted "compulsion" and "regimentation," he said that farmers were free to ignore their allotments. But only those who complied would be eligible for maximum benefit payments under the soil conservation program and for corn loans.

When Andy Hodges, chairman of the McDonough county corn control committee (McDonough was the county in which Macomb was located), was asked if his group would be holding any more meetings. He said they had already held a series of meetings and "won't hold any more."[89]

In a later statement Wickard said "individual inequities will be adjusted but the administration will not give an inch" to demands for a general increase in corn acreage allotments. The allotments he said had been made according to the law, and the only way they could be changed was by congressional amendment. So the farmers could take them or leave them.[90]

Earl Smith, of the National Farm Bureau and the Illinois Agricultural Association, warned critics of the farm program: "Should our opponents succeed in their objective of dividing farmers' opinion and cooperation, they will carry a tremendous responsibility within a few months, for the result of such efforts will be to drive America still deeper into business stagnation and depression."[91]

But later the Secretary of Agriculture attempted to still the protests by modifying the deductions from base benefits for overplanting some crops, such as corn or wheat. Deductions of 5/8 of the schedule originally announced would be permitted. The Department believed that this modification would reduce much of the objections that had been raised to the agricultural control program.[92]

On May 13 officials of the Agricultural Administration advised farmers that if they refused to cooperate with the voluntary corn acreage reduction program, they would probably face compulsory control measures at harvest time. Speaking over the radio on the 12th, Wallace told corn growers that if they did not cooperate, they faced the prospect of "being swamped by wasteful and devastating overproduction, painfully low corn prices," and reduced buying power. He also said that they should follow the example of southern cotton and tobacco growers who had courageously voted to limit the sales of their products by using marketing quotas.[93]

However, newspaper reports a few days earlier told a different story. According to these reports, the Senate had voted to adopt a series of amendments to the Farm Act after Southern Democratic senators told of widespread resentment against the cotton and tobacco quotas. Southern farmers had voted for compulsory control without knowing what their quotas would be. Once they received their quotas they too were holding mass protest rallies.[94]

Other newspapers also attacked the Farm Act and its administration under the Department of Agriculture. As news of the revolt spread, William Hirth of the Missouri Farmers Association commented in an editorial, "At last a real revolt is coming into being against the un–American policies of Secretary Wallace, and the writer wishes this movement Godspeed. If the late tawny-haired Milo Reno of Iowa were still among the living, how he would welcome the spirit of protest which is now arising in the land against Wallace, whom nobody knew better than Reno, or despised more cordially."[95]

The *Daily Journal* of Macomb, Illinois, carried an editorial from a newspaper in Kansas. A farmer complained: "I have lived a long time in Kansas and never before have I seen so many inspectors running around investigating so many different kinds of things and butting into everybody's private business as are infesting the country now." The writer of the editorial concluded that in the United States the average producer was carrying on his back one-eighth of an office holder who "in some form or other ... is directing his movements."[96]

Four corn belt farm dailies attacked the Federal government's crop control program in late July, charging, "The government has experimented with cotton, peanuts, tobacco, wheat and corn and hogs. The most marked result has been not the adjustment of production to demand but maladjustment between crops."[97]

In late July the Agricultural Adjustment Administration officials announced that an expansion in domestic markets and prospects for increasingly larger exports of corn might make corn marketing quotas unnecessary.

According to the officials, the first marketing quota estimates had been based on exports of 25,000,000 bushels of corn annually, but shipments abroad since October of 1937 had been estimated at more than 105,000,000 bushels and might total more than 150,000,000 bushels before the new corn marketing year began in October 1938. One of the main factors in the increased demand for corn in the foreign markets was a small corn crop in Argentina, a major competitor of the United States in the exportation and sale of corn.

Estimates of domestic consumption also had been rising. Livestock reports of hog supplies indicated the need for more corn in the domestic market than had first been estimated.

There was some suspicion that the change in plans was made because of dissatisfaction in the corn belt and the activities of the Corn Belt Liberty League. It may be that Wallace was afraid to put his marketing quota to the test. "At least he has revised his figures until a vote is not considered necessary."[98]

Many people, even some county agents, said that if there had been a corn quota vote called, it would not have passed in many localities. One observer said that he saw disaffection in evidence everywhere in the Middle West, and county agents were admitting that they had lost control of the situation.[99]

The leaders of the Corn Belt Liberty League of Macomb, Illinois, sent out a letter to members and friends thanking them for "their great help in warding off the welding of one of the links in the chain of bondage that the AAA and others planned for agriculture — the referendum on corn marketing quotas.... We feel we are not egotistical in claiming some part in this accomplishment."[100]

After the corn quota referendum was called off, a farmer in Minnesota wrote Casement that he expected to have 4000 farmers in his county united in an organization against government crop control. "As you say, we have got these Government Dictators on the run and we want to keep chasing them until they mind their own business instead of telling us how to run our farms."[101]

The wheat program was still a concern in Kansas and states further west. Thomas Doran wrote Casement that he had heard from President Farrell of the Agricultural College that only about 40 percent of the farmers who signed up in 1937 were signing up in 1938. Farrell reported that "the farmers are totally disgusted with the program and cannot be convinced."

Doran said that he had heard that he could not even sow wheat for pasture, even though he had "never signed up with them, have never taken any of their tainted money and have no use for their programs."[102]

Others also reported their disgust with the government wheat reduction program. Some disliked the program because they genuinely disliked governmental interference. Others felt that the government was asking them to do too much to obtain their payments. Through eastern Kansas and western Missouri the Department of Agriculture deductions were estimated to fall between 30 and 50 percent. This was due to recent changes in area crops from corn to wheat because of the recent drought, which had destroyed corn crops.

A farmer in Cass County, Missouri, said that a reduction of 50 percent would be very difficult for him. "Farmers have purchased wheat machinery and its investment can be returned only through growing wheat. Our meadows have been destroyed by drought, and livestock herds depleted through lack of feed. What are we to grow the coming year unless wheat? Even should we turn to more diversified farming with increased livestock, it would be a year or more before we began to get income from the change in operations, and meantime our need for cash to pay overhead continues.... It appears to me the big landowner with capital behind him, or the large owners of mortgaged farms, can make the switch and qualify more easily for the benefit payments than can the average farmer trying to make a living on 160 to 240 acres."[103]

Other reports mentioned that politics might be influenced by the farm revolt. "The human element of farmer cooperation, in a year which finds politics rampant in the farm lands may prove the final test of the (farm) program."[104]

The strength of farmer discontent with the farm program was a matter of much speculation. Hopeful Republicans looking toward the November elections predicted that the storm of protests would assume hurricane proportions and "blow the Democrats out of office." Democrats admitted that there was some discontent, but minimized its importance.[105]

Secretary Wallace came to Springfield, Illinois, on October 14 to defend the aims of the program. He said that the farmer's cash income this year would only be about 12 percent less than last year. He contrasted that with factory employment and payrolls, which had gone down more than 30 percent. The AAA program of crop control was responsible, he said, for stabilizing the return to the farmer.

He said that the act was capable of improvement, but that improvements should be made using the program that had already been proven worthwhile. "Because the basic structure has been built strong and well."[106] Wallace felt that the Liberty League in Illinois had made charges last spring that "turned out to be exactly 100 percent bunk."[107]

As the election campaign of 1938 wound down to a close, newspaper commentators reviewed the campaign and tried to decide what the most important issues had been. A *New York Times* correspondent commented that he believed this election would give the answer, in part at least, "to the question of whether the country wants the New Deal to become a permanent order of things." Taking a closer view at several important states, this commentator said that in some of the states an issue of national significance was the discontent of the farmers with the New Deal farm program. And, of course, Republican hopes for revival of their party rested on the election.[108]

The election was held on Tuesday, November 8. By Wednesday the newspapers told the story. The Republicans had made big gains in the 1938 elections. For the first time in six years Democrats across the country either experienced close finishes with their rivals or the unfamiliar taste of defeat. Republicans gained six or more Senate seats and fifty or more seats in the House of Representatives.[109]

In the 1938 Senate election in Kansas, Senator George McGill, one of the sponsors of the Agriculture Control Bill, was defeated by Republican Clyde Reed. In North Dakota Gerald Ney, a Republican, was elected; in Ohio Robert Taft, a Republican, was elected. In South Dakota Chandler Gurney, a Republican, was elected; in Wisconsin Alexander Wiley, a Republican, was elected. In Iowa Senator Guy Gillette — a Democrat, but one whom Roosevelt had attempted to purge — was reelected.[110]

In the 1938 elections to the House of Representatives, 10 Republicans were elected in Illinois, seven in Indiana, seven in Iowa, six in Kansas, 12 in Michigan, seven in Minnesota, three in Nebraska, 14 in Ohio, two in South Dakota, and eight in Wisconsin. In many of these states Republican Representatives predominated.[111]

With these new Republicans in the House and Senate, and the conservative southern Democrats already in place in both houses, many people predicted that Congress would turn away from the trends of the New Deal.

Following the election, Tilden Burg, president of the Corn Belt Liberty League, was jubilant. He said that it was "proof that we, the American people, intend and will subtract from their authority, or entirely eliminate as officers any and all who misuse the trust placed in them as our country leaders and who desire to take away instead of adding to the freedom of the American people."

He said that their protest had not been partisan. The Corn Belt Liberty League had tried to avoid any political entanglements. It was for that reason that they did not encourage any meetings in the last two months before the elections. Their purpose had been to eliminate the compulsory features of the Agricultural Control Act."[112]

Dan Casement was naturally gratified at the results of the Kansas elections. He received several telegrams congratulating him for the work he had put into achieving these results. Kurt Grunwald, a friend from the Farmers Independence Council days, wired: "DUE TO YOUR EFFORTS KANSAS HAS RETURNED TO SANITY CONGRATULATIONS." Alex Smith wrote: "HEARTIEST CONGRATULATIONS ON WONDERFUL RESULTS IN KANSAS. MORE POWER TO YOUR ELBOW."[113]

Stanley Morse wrote a jubilant letter to former president Herbert Hoover: "For the first time in six years we shall have a better reason than usual for giving thanks to the Almighty next week. This reason includes your splendid leadership."[114]

Hoover replied, "I am wondering whether or not you thought it might be worthwhile just to make a hint to some group of farmers that they might reconsider my program of 1931 which rested upon the government's leasing for ten or twenty years the submarginal lands to whatever extent was necessary to get agriculture in balance.... I was convinced then, and I am convinced now, that this program would not cost one-third as much as the present program...."[115]

For those who had fought the New Deal farm program it looked as if they had won a victory. They expected that changes would be made because of the defeat of some of the supporters of the New Deal farm programs and the election of greater numbers of Republicans to Congress.[116]

Later, James Farley, Chairman of the Democratic National Committee, Roosevelt's political advisor and head of the United States Post Office, wrote to Democrats around the country asking them why the party had suffered such defeats in the 1938 election. Many of the replies blamed farm resentment for the switch in voting patterns. Governor Frank Murphy of Michigan wrote, "Undoubtedly the drop in farm prices—beginning in mid-summer of 1937 and reaching a low point in the mid-summer of 1938—played a considerable part in the disaffection of the voting public which since 1932 has been preponderantly Democratic. Many farmers, and doubtless some wage-earners, registered a protest vote against the decline in their purchasing power."[117]

Iowa Democratic State Senator Dan Mason, who was defeated in the

1938 election, wrote to President Roosevelt: " I think Mr. President we farmers will have to be considered, all we want is a fair price."[118]

Despite their dreams that their protest would effect major changes in the Administration's agricultural programs, this did not materialize. Much of the program remained the same. And the leaders of the Corn Belt Liberty League returned to their farming and do not seem to have made any other forays into the world of political action. Dan Casement also remained on his ranch in Kansas, dying at the age of 88 in the early 1950s.

10

Conclusion

The preceding chapters have told the story of several farm organizations and their criticism of the New Deal. Many of the groups initially supported Roosevelt in the 1932 campaign. But their critique began soon after his election, continued through the formulation and passage of the Agricultural Adjustment Act, and ended some time in 1942, when their leaders gave up hope.

The groups had begun with high hopes. The Farm Holiday attracted the most attention, but actual membership totals were never revealed. In February 1933 Reno wrote Hirth that they had 90,000 members, but this is probably an exaggeration. Actual dues-paying members may have been under 10,000.[1] The Missouri Farmers Association had 18,000 members, and its journal, *The Missouri Farmer*, reached 50,000. The Farmers Independence Council was very small, with perhaps 400 members, and supported more by wealthy East Coast backers than by farmers or ranchers. The National Farmers Process Tax Recovery Association may have had 4,000 members, but they hoped their group would grow to include 20,000 or more members. In 1937 Donald Van Vleet wrote:

> We have in Iowa around two thousand members and also at the present time have a national organization underway in ten other states, and before Congress is in session we believe we will have at least twenty thousand members.[2]

Throughout the nation there were hundreds of thousands of farmers who might be expected to have joined up with these organizations. But they did not join. The groups were not successful either in attracting a large membership or in defeating the New Deal agricultural program.

In the end the reader is left with a question: Why did the groups fail in achieving their goals? They had enthusiasm, leaders of some political

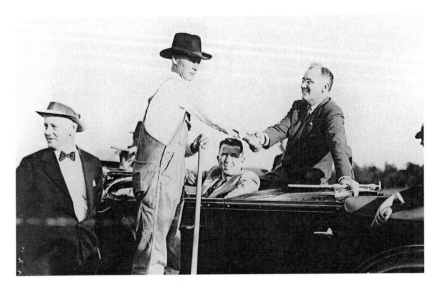

President Roosevelt greeting a farmer on his way to Warm Springs, Georgia.
(Courtesy of Franklin D. Roosevelt Library, Hyde Park, N.Y.)

experience, and an amount of political power. Their organizations spread into 10 or 11 middle western states. But they failed. What were the reasons behind their failures?

Let us look again at the groups and their situations. Each group initially had enthusiasm. Their enthusiasm was based on several things. Many had ties to the farm organizations of the 19th century, to the Grange, Farmers Alliances or the Populist Movement. Some were fueled by radical Farmers Union leaders who had opposed the Agricultural Adjustment Act from the very beginning because it was not the cost of production plan that they had developed and supported through the Farmers Union. The Farmers Independence Council, while led by a philosophy less receptive to governmental involvement than many of the other groups, also attempted to attract any farmers who opposed the New Deal.

The groups had strong leaders. John Simpson led the Farmers Union in the early Depression years. William Hirth organized the Missouri Farmers Association through small farm clubs located throughout the state. Milo Reno led the Farm Holiday and suggested the organization of the National Farmers Process Tax Recovery Association. Edward Kennedy was very active in the Farmers Union, Farm Holiday and the Recovery Association, and formed the Farmers Guild. These men all had been working in the Farmers Union since the early 1920s. Again Dan Casement of the Farmers Independence Council was an exception. He had remained outside of these organizations but had been active in ranchers associations, and also was known as a writer.

Political infighting is part of this story. Simpson, Reno and Kennedy ousted some of the leaders of the Farmers Union — such as Huff, Ricker, Thatcher and Talbott — from power in 1930. These leaders did not leave the Farmers Union but instead waited their time and regained control of the Farmers Union in 1936 after the deaths of Simpson and Reno. They then showed their distaste for Kennedy, as a supporter of Reno and radical ideas (such as cost of production, return of the processing tax, and support of third party candidate William Lemke), by defeating Kennedy in his bid for reelection to the office of National Secretary that same year. Following the defeat of Kennedy, the small groups (such as the National Farmers Process Tax Recovery Association) who were trying to regain their processing taxes could not be sure of receiving support from the Farmers Union.

They became, instead, the rump faction, the group who were attached to that part which finally pulled out of the Farmers Union. (Although, significantly, Iowa did not join the Farmers Guild.) The National Farmers Process Tax Recovery Association and the Farmers Guild often suggested that they were part of the true Farmers Union, holding the older ideas of the group. They had the leadership of Kennedy and Lemke, who were unhappy with the Farmers Union.

The ties with Edward E. Kennedy and William Lemke gave many of these groups both strengths and weaknesses. These two men were experienced and capable, and worked very hard to get the bills that they favored across. Kennedy had friends in the Farmers Union in the Middle West. Many state Farmers Union presidents became leaders in the National Farmers Process Tax Recovery Association. But both Kennedy and Lemke also had powerful enemies within the Farmers Union. At one time, when Milo Reno and John Simpson were alive, they had been aligned with the dominant leaders of the organization. But these two leaders had died. And the new powers in the Farmers Union disliked Kennedy and Lemke. They wanted to turn from opposition to FDR to support of the President and his programs. Kennedy lost his fight to maintain his position as secretary of the Farmers Union in 1936, defeated by the leaders (many of whom came from the western states where they were more powerful).

The groups did not really belong to any political party. Many of the Farmers Union leaders had voted for Roosevelt in 1932, but became disenchanted with the Roosevelt Administration because of the Administration's agricultural programs. Johnson, Kennedy and Lemke were prime examples of this trend. The farm groups did not receive strong support from either major political party. The groups did have the support of some of the congressmen from local states: Frazier and Lemke of North Dakota, Gurney from South Dakota, Gilchrist and Gillette of Iowa. The only Democrat of this group is Gillette, who had antagonized Roosevelt by refusing to support his court packing plan. In response, Roosevelt and some members of his cabinet had tried to purge Gillette from the Democratic ticket, a move that failed. Some of the members of the Farmers Union, and the National Farmers Process Tax Recovery Association, were associated with the Union Party. Kennedy's association with the Union Party was especially strong. Generally, the leadership of the groups voted Republican because they opposed the New Deal.

The Farmers Independence Council had some ties to the Republican Party. As funds for the group dried up, several of the organizers of the group went to work in the 1936 Republican campaign.

The groups were enthusiastic because they believed that they were right. However, their views on what they thought the federal government ought to do in the Depression crisis differed. The Farmers Independence Council did not believe that government should regulate farmers or ranchers at all. The Farmers Union, Farm Holiday and Farmers Guild members believed that federal regulation should proceed along the lines of their ideas of cost of production, i.e. establishing prices for farm commodities and requiring that all goods be bought at these prices.

The National Farmers Process Tax Recovery Association believed they had a right to the hog processing taxes. They believed this because they had seen the prices of hogs go down following the imposition of the processing tax. They believed it because the packers told them that they were paying them less because of the tax. (Quite possibly, the packers were trying to encourage the farmers in 1933, 1934 and 1935 to protest against the tax, as that might increase the pressure on Congress to abolish it.) The farmers believed it because the leaders of the National Farmers Process Tax Recovery Association were able to obtain documents from the Department of Agriculture showing that economists in the Bureau of Economics of the Department of Agriculture had said that the taxes were not paid by the packers nor by the consumers. If these two groups did not pay the tax, the farmers reasoned they should be able to prove that they had paid the tax. But they could not produce the receipts that the courts required to show that they had paid the tax.

The Corn Belt Liberty League protested the proposals of the Department of Agriculture that would require all farmers to follow departmental regulations in the type and amount of goods produced. All felt that the New Deal farm program was not working, showed favoritism, and was an intrusion on traditional agrarian independence.

Why did the groups not grow? Why did so few farmers join the group? The Missouri Farmers Association was fortunate in having an established farm journal to spread the ideas of William Hirth and other leaders in the Association. However, many of the groups had difficulty in getting their message across to the average farmer. Milo Reno struggled to get the funds to keep a paper going to inform the members of his group of current activities. Gurney was able to broadcast his message over the radio station which he owned, WNAX, but he was involved in a struggle to keep the station and eventually lost it. Lemke was limited in his access to the farmers. During the campaign of 1936 he campaigned across the nation, but after that his access to the farmers was limited to columns in a few newspapers, such as the *Unionist and Public Forum*, and occasional radio speeches. Casement and the Farmers Independence Council were often turned down in their requests for radio time or given times when it was not likely that many people would be listening.

Kennedy had been editor of the *National Farmers Union* paper until his ouster from his position as secretary of the Farmers Union in 1936. After that, he published his own newsletter, the *Washington News Letter*, but that mainly reached the believers. Kennedy, too, was limited in his access to state Farmers Union papers. His newsletter was often printed in the *Iowa Union Farmer*, as were many articles about the National Farmers Process Tax Recovery Activities.

But the same pattern was not followed in neighboring states. In fact, several state Farmers Union papers printed the Department of Agriculture Martin White statement warning farmers to stay away from groups who might be fraudulently trying to get the farmers' money in tax recovery schemes.

Farm Holiday, Corn Belt Liberty League and Recovery Association local recruiters often complained of the difficulty they experienced in getting their notices of meetings and tax recovery information in the papers. They said the papers were more sympathetic to the Farm Bureau cause and would not publish their materials.

An Indiana hog farmer wrote Congressman Lemke that he had conducted his own poll. He asked farmers at the Chicago, Cincinnati and Indianapolis stockyards what they thought about the Frazier-Lemke Hog Processing Tax Refund Bill. "They thought you were doing them justice and thought they were entitled to it. But, Mr. Lemke, not one in twenty-five knew about it." He said he subscribed to several state newspapers and had never seen anything in these papers about the attempts to recover the hog processing tax for the farmers.[3]

This narrowing access to farm papers may have been due to the growing friendship of some of the Farmers Union officials to the Department of Agriculture. Lemke and Kennedy said the leaders of the Farmers Union were too friendly with the Department of Agriculture. They said that these men had not "stood solidly behind any real legislation. Some of them have rather been the chore boys of the Secretary of Agriculture."[4]

Finally, the Department of Agriculture opposed the programs of these farm groups. This is shown in the case of the National Farmers Process Tax Recovery Association. When the question of refunding cotton or tobacco or potato taxes came up, the Department of Agriculture offered only mild protests. And these refunds were passed in Congress. But repeatedly the Department of Agriculture worked to defeat the passage of the hog processing tax refund. Sometimes strong letters were written by members of the Department of Agriculture or other departments. Tactics were employed that Kennedy and Lemke (at least) felt were directed by the Department of Agriculture to defeat their legislation. Sometimes farm leaders seemed to have been influenced by government jobs or favors.

County extension agents, influenced by the Farm Bureau and the Department of Agriculture, often seemed to oppose the National Farmers Process Tax Recovery Association. They advised their farmers against joining the Recovery Association. Sometimes they withheld records from farmers who wished to secure them for the purposes of establishing a claim with the Association. They repeatedly told the farmers that they did not need to join any organization because if the federal government wanted to return their taxes they did not need any "go-between."

The large packers refused, in many instances, to give the farmers their records. Farmers may have despaired of their ability to prove that they had sold hogs during the years when the processing taxes were in effect. Now that the taxes were no longer in effect, the packers were not as anxious to assure the farmers that they had paid the taxes. In fact, during most of this time the packers were trying to keep as much of the processing taxes as they could for themselves.

So the farmers received conflicting advice from various quarters as to whether they should join the organizations or not. And the groups remained small and ineffectual.

There seemed no way out. In Congress they could not produce the votes that would ensure passage of their legislation. The reason they could not produce the votes was because their groups had remained small.

Also, there is a philosophical inconsistency involved in the activities of the organizations. For example, the National Farmers Process Tax Recovery Association was founded initially to get back the process tax for farmers who had tried to avoid being organized by the Agriculture

Adjustment Administration. Thus, farmers who had originally resisted being organized were now urged to join another organization. Also, this organization was expensive—at a time when farmers had very little money.

Perhaps it is not so hard to see why most of the organizations did not grow. As their group's fortunes waned, their members scattered in different directions. Simpson and Reno died in the 1930s. Donald Van Vleet, the group's first president, who quit in a disagreement with Kennedy, went on to become first a seller of cooperative products for the Iowa Farmers Union and then, by 1942, the president of the Iowa Farmers Union.[5] Edward E. Kennedy became a research director and economist for the United Mine Workers of America, and later a probate judge for the state of Maryland.[6] A.J. Johnson remained on the farm in Moorhead, Iowa, but served in the winters as a Republican legislator in the Iowa State legislature for several terms.[7] William Lemke lost his fight for election to the U.S. Senate in 1940 but returned to the U.S. House of Representatives in 1942, where he remained in office until his death in 1950.[8] D.B. Gurney died in the early 1940s.[9] Dan Casement and William Hirth died in the 1940s. The farmers who belonged to the groups these men led turned to other interests. Only the Missouri Farmers Association lasted past the war years.

Perhaps, more than anything else, the war finally finished off most of these groups. People's attentions were drawn in other directions. Some farmers became more affluent with the rising prices of agricultural products.[10] They had less need for the groups. Other farmers left the countryside and found jobs for themselves in the booming war industries in the cities.

In a sense, the farm critics had been trying to fulfill an old, old dream of small independent farmers who could influence government legislation. This had been the rallying cry of John Simpson, Milo Reno, Dan Casement, William Hirth and others. But these agrarian radicals were all old men, and by 1942 they had disappeared from the national scene.

The New Deal had begun the change. And some of the old radicals had fought it. (That was the reasoning behind much of their criticism.) World War II completed the change as more and more of the small independent farmers left the land.

Still, these old letters and records of the various critics of the New Deal remain to remind us of people's concerns in an earlier day—concerns that are echoed in the rallying cries of farmers' protests of more recent years.

Notes

Chapter 1

1. Reminiscing in the Rush Hill Area, Centennial Yearbook, 1981, pp. 28–29. William Hirth, "A Plea for Little Bob White," *The Missouri Farmer*, May 15, 1932, p. 1.

2. William Hirth, "Foreword," *The Missouri Farmer and Breeder*, October 15, 1908, p. 1.

3. William Hirth, "The Great Weakness of Agriculture," *The Missouri Farmer*, February 15, 1933, p. 7.

4. Laurence Goodwyn, *The Populist Moment, a Short History of the Agrarian Revolt in America*, Oxford: Oxford University Press, 1978, p. 179.

5. Ray Derr, *Missouri Farmers in Action*, Columbia: Missouri Farmers Press, 1958, p. 27.

6. Interview with Ray Derr, *Missouri Farmers in Action*, p. 33.

7. William Raymond Young, officer in the MFA, July 7, 1992. Derr, "Hirth, What Is Really the Matter with Agriculture?" *The Missouri Farmer*, March 15, 1930, p. 1.

8. Derr, *The Missouri Farmers in Action*, p. 36.

9. William Hirth, "A Foreword," *The Missouri Farmer and Breeder*.

10. Interview with Raymond Young, MFA officer and friend of William Hirth, Columbia, Missouri, July 7, 1992.

11. William Hirth, *The Missouri Farmer*, February 1914, p. 1.

12. Aaron Bachtel, *The Missouri Farmer*, June 15, 1933, p. 7. Derr, *Missouri Farmers in Action*, p. 39. Chuck Lay, "He Loved Agriculture and All Things Beautiful," *Today's Farmer*, October 1990, pp. 28–29.

13. Interview with Raymond Young.

14. Interview with Raymond Young.

15. Theodore Saloutos and John D. Hicks, *Twentieth Century Populism*, Lincoln: University of Nebraska Press, 1951, pp. 57–58.

16. Gilbert Fite, *American Farmers, The New Minority*, Bloomington: Indiana University Press, pp. 16–17.

17. William Hirth, "Speech" Quoted in Derr, *Missouri Farmers in Action*, p. 43.

18. Frank F. Ross, "A Farm Club at Work," *The Missouri Farmer*, March 1, 1916, p. 8. Derr, *Missouri Farmers in Action*, p. 44.

19. Hirth, "Lafayette County Points the Way," *The Missouri Farmer*, October 1, 1916, p. 1.

20. Hirth, "Farm Clubs Forming Everywhere," *The Missouri Farmer*, January 1, 1916, p. 8.

21. Derr, *Missouri Farmers in Action*, pp. 44–45.

22. Hirth, "Are You Satisfied?" *The Missouri Farmer*, September 1, 1916, p. 1.

23. "Farm Club Saves $200 on Bulk Twine," *The Missouri Farmer*, June 11, 1916, p. 12.

24. William Hirth, quoted in Derr, *Missouri Farmers in Action*, p. 47.

25. Derr, *Missouri Farmers in Action*, p. 48.

26. Raymond Young interview.

27. William Hirth, quoted in Derr, *Missouri Farmers in Action*, p. 50.

28. William Hirth, quoted in Derr, *Missouri Farmers in Action*, p. 51.

29. Derr, *Missouri Farmers in Action*, pp. 61–62. Interview with Raymond Young.

30. William Hirth, "Just Suppose?" *The Missouri Farmer*, September 1, 1919, p. 16.

31. William Hirth, "The American Market for the American Farmer," *The Missouri Farmer*, September 1, 1919, p. 16.

32. United States Department of Agriculture, Agricultural Statistics, 1936. Washington: U.S. Government Printing Office, 1936.

33. South Dakota Crop and Livestock Reporting Service, South Dakota Agriculture, Sioux Falls, SD, 1942, p. 15.

34. *Ibid.*, p. 390.

35. Department of Agriculture, Agricultural Statistics, 1937, p. 410.

36. William Hirth to Xenophon P. Wilfley, February 4, 1926. Hirth Papers.

37. Gilbert C. Fite, *George N. Peek and the Fight for Farm Parity*, Norman, Oklahoma: University of Oklahoma Press, 1954, p. 40.

38. Theodore Saloutos and John D. Hicks, *Twentieth Century Populism*, Lincoln: University of Nebraska Press, 1951, pp. 380–381.

39. Saloutos and Hicks, *George N. Peek and the Fight for Farm Parity*, pp. 97, 98.

40. William Hirth to Clarence Cannon, March 27, 1924, Davis Papers, Western Historical Manuscript Collection, University of Missouri.

41. Lowell K. Dyson, *Farmers Organizations*, Westport, CT: Greenwood Press, 1986, pp. 60–63.

42. Saloutos and Hicks, *Twentieth-Century Populism*, p. 386.

43. William Hirth, "Voice of a Million Farmers, Corn Belt Committee Meets at Des Moines," *The Missouri Farmer*, August 15, 1925, p. 7.

44. "Minutes of the Corn Belt Federation," May 17–18, 1927. Western Historical Collections, University of Colorado at Boulder.

45. Fite, *George N. Peek*, p. 153. Derr, *Missouri Farmers in Action*, pp. 80–81.

46. William Hirth to John Napier Dyer, February 6, 1929. Hirth Papers.

47. Derr, *Missouri Farmers in Action*, pp. 64, "The Big Convention," *The Missouri Farmer*, September 1, 1925, p. 10.

48. T.H. De Witt to S.B. Elting, March 6, 1928; Frank N. Scott to William Hirth, March 14, 1928; Ruby Hulen to William Hirth, May 15, 1928; T.H. DeWitt to all club secretaries, August 15, 1928. Hirth Papers.

49. "Tenth Annual M.F.A. Convention a Rousing Success," *The Missouri Farmer*, September 15, 1926, pp. 1, 7.

50. "Convention Comment," *The Missouri Farmer*, October 1, 1928, p. 18.

51. A.R. Lappell to William Hirth, October 16, 1928; Wm. Bret to William Hirth, January 11, 1929; Oscar Royce to William Hirth, January 25, 1929; William Hirth to Oscar Royce, January 31, 1929; William Hirth to Edwin Philipps, March 11, 1929. Hirth Papers.

52. Theodore Saloutos, "William A. Hirth: Middle Western Agrarian," *Mississippi Valley Historical Review*, p. 28, September 1951, pp. 215–232; William Hirth to Charles McNary, April 23, 1927. Davis Papers, Western Historical Manuscript Collection, Columbia, Missouri.

53. Hirth, "Governor Smith Defines Position on Farm Relief," *The Missouri Farmer*, October 1, 1928, pp. 1, 6.

54. William Hirth to George W. Norris, April 22, 1929. Hirth Papers.

55. Hirth, "Highlights of the Great Convention," *The Missouri Farmer*, September 1, 1930, p. 8.

56. Hirth, "The Farmers' Living Standard," *The Missouri Farmer*, September 15, 1929, p. 10.

57. Hirth, "Keeping Out of Politics," *The Missouri Farmer*, May 15, 1931, p. 8.

58. William Hirth to Chas. G. Dawes, May 5, 1932. Hirth Papers, Western Historical Manuscript Collection, University of Missouri, Columbia, Missouri.

59. William Hirth to Judge Xenophon Caverno, February 26, 1932. Hirth Papers, Western History Collection, University of Missouri, Columbia, Missouri.

60. Hirth, "Squaring Away for the Big Presidential Contest," *The Missouri Farmer*, March 1932, p. 1.

61. William Hirth to James A. Farley, June 11, 1932. Hirth Papers, Western Historical Collection, University of Missouri, Columbia, Missouri.

62. Hirth, "A Victory for the Forgotten Man," *The Missouri Farmer*, July 15, 1932, p. 1.

63. Richard O. Davies, "The Politics of Desperation: William A. Hirth and the Presidential Election of 1932," *Agricultural History*, 389 (1964), p. 226.

64. "Resolutions of the M.F.A. Convention," *The Missouri Farmer*, September 15, 1932, p. 6.

65. William Hirth to Franklin D. Roosevelt, November 8, 1932. Presidential Papers, Franklin D. Roosevelt Library, Hyde Park, New York.

66. Franklin D. Roosevelt to William Hirth, January 17, 1933. Presidential Papers, Franklin D. Roosevelt Library, Hyde Park, NY.

67. William Hirth to James M. Thompson, November 9, 1932. Hirth Papers.

68. William Hirth to Franklin Roosevelt, November 14, 1932. Hirth Papers.

69. William Hirth to James Farley, Nov. 8, 1932; William Hirth to Louis Howe, Dec. 21, 1932. Hirth Papers.

70. Saloutos and Hicks, *Twentieth Century Populism*, pp. 460–461.

71. William Hirth to Clarence Cannon, November 28, 1932. Hirth Papers.

72. William Hirth to J. Caverno, November 18, 1932. Hirth Papers.

73. Hirth, "An Unsound Farm Proposal," *The Missouri Farmer*, April 1, 1933, pp. 3–4.

74. "Farm Wages Downward," *The Missouri Farmer*, May 1, 1933, p. 16.

75. "Flour Bag Aprons," *The Missouri Farmer*, September 15, 1933, p. 11. "Dyeing Flour Bags Easily," *The Missouri Farmer*, May 15, 1933, p. 11.

76. Capitola Gradwohl, "Birth Control on the Farm," *The Missouri Farmer*, May 15, 1934.

77. "The Agricultural Adjustment to Its Logical Conclusion," *The Missouri Farmer*, May 15, 1934, p. 5.

78. Hirth, "A New and Strange Philosophy," *The Missouri Farmer*, January 15, 1934, p. 1. Theodore Saloutos, *The American Farmer and the New Deal*, Ames: Iowa State University Press, 1982, pp. xv–xvi.

79. Hirth, "Are You Doing Your Share?" *The Missouri Farmer*, May 15, 1916, p. 1.

Chapter 2

1. John L. Shover, *Cornbelt Rebellion: The Farmers Holiday Movement*, Urbana: University of Illinois Press, 1965, p. 28. Frank Dileva, "Iowa Farm Price Revolt," *Annals of Iowa*, January 1954, p. 32, 172.

2. J.H. Cappy to Elmer Thomas, November 21, 1931. Elmer Thomas Collection, Carl Albert Center Congressional Archives, Norman, Oklahoma.

3. Geo. G. Walker to Elmer Thomas, November 26, 1931. Elmer Thomas Collection.

4. W.A. Stone to Elmer Thomas, November 9, 1931. Elmer Thomas Collection.

5. Charles M. Evans to Elmer Thomas, December 1, 1931. Elmer Thomas Collection.

6. "John Andrew Simpson," *Dictionary of American Biography*, New York: Charles Scribners Sons, 1987, pp. 180–181.

7. Gilbert A. Fite, "John A. Simpson: The Southwest's Militant Farm Leader," *Mississippi Valley Historical Review*, pp. 35, 563–567.

8. A.W. Ricker, "The Birth and Growth of the Northwest Farmers Union," *Farmers Union Herald*, Oct 1937, pp. 1–2.

9. Theodore Saloutos, *The American Farmer and the New Deal*, Ames, Iowa: Iowa State University Press, 1982, pp. 19–20.

10. Lowell K. Dyson, "National Farmers Union," p. 219. Charles Anthony Mast, "Farm Factionalism over federal agricultural policy: The National Farmers Union, 1926–1937." M.A. Thesis, University of Maryland, 1967, pp. 17–23.

11. Dyson, "National Farmers Union," p. 220.

12. J.E. McClendon to Elmer Thomas, November 28, 1931. Elmer Thomas Collection.

13. Odes Harwood to Elmer Thomas, November 18, 1931. Elmer Thomas Collection.

14. Geo. Walters to Elmer Thomas, November 26, 1931.

15. "Simpson New National President," *Nebraska Union Farmer*, November 26, 1930, p. 1.

16. Theodore Saloutos and John D. Hicks, *Twentieth Century Populism*, Lincoln: University of Nebraska Press, 1951, p. 405.

17. *Ibid.*, pp. 405–409.

18. Dyson, "National Farmers Union," p. 220. Saloutos, *Farmer Movements in the South*, pp. 272–273.

19. Federal Farm Board, News Release, November 19, 1930. National Farmers Union Collection, Historical Collections, University of Colorado at Boulder.

20. C.E. Huff to Edwin Aschenbrenner, April 30, 1956. National Farmers Union Collection.

21. *Ibid.*

22. Saloutos and Hicks, *Twentieth Century Populism*, pp. 413, 419–420.

23. *Pioneer Press*, January 16, 1930, quoted in Saloutos and Hicks, *Twentieth Century Populism*, pp. 419–420.

24. Saloutos and Hicks, *Twentieth Century Populism*, p. 421.

25. *Nebraska Union Farmer* (Omaha), August 14, 1928. Quoted in Saloutos and Hicks, *Twentieth Century Populism*, p. 429.

26. John Simpson, "Report of the National President," Minutes of the Twenty-Seventh National Convention, Farmers' Educational and Cooperative Union of America, November 17–18, 1931, p. 14.

27. Elmer Thomas to John Simpson, November 26, 1930. Elmer Thomas Collection.

28. "Legge Calls Critic 'Unmitigated Liar.'" Elmer Thomas Collection.

29. John Simpson to Alexander Legge, December 24, 1930. Elmer Thomas Collection.

30. Milo Reno to Charles McNary, December 23, 1930. Elmer Thomas Collection.

31. "Letter from Legge Quoted, McNary Makes Public Epistle from Farm Board Chairman." Elmer Thomas Collection.

32. "Simpson Exposes Farm Board," *Oklahoma Union Farmer*, January 1, 1931, pp. 1, 3.

33. Saloutos, *The American Farmer and the New Deal*, p. 30.

34. "Critics of the Farm Board Lack Vision, Chairman Charges," *Farmers Union Herald*, May 4, 1931, p. 1

35. "Debunking Bunk" Radio Address, February 24, 1934. *Oklahoma Union Farmer*, Elmer Thomas Collection. And in Denmark, where Simpson and his wife also visited, farmers, although about the same percentage in the United States, "dominate all government affairs…. In the United States, through lack of organization, farmers have very little to say about the affairs of our state and national governments." John Simpson, "Cooperation," *The Militant Voice of Agriculture*, pp. 29–32.

36. "What's the Matter? What's the Cause? What's the Remedy?" Address by John A. Simpson, *Michigan Producer*, October 10, 1931.

37. "Senators Are Given Vivid Example of How Middlemen Exploit Producers and Users," *Colorado Union Farmer*, December 1931. Simpson Collection.

38. T.E. Howard to John Simpson, December 28, 1931. John Simpson Collection.

39. John Simpson to T.E. Howard, January 5, 1932. John Simpson Collection.

40. R. Douglas Hurt, *The Dust Bowl: An Agricultural and Social History*, Chicago: Nelson-Hall, 1981, p. 33.

41. Franklin Roosevelt to John Simpson, March 7, 1932. John Simpson Collection.

42. John Simpson to Franklin Roosevelt, April 11, 1932. John Simpson Collection.

43. Franklin Roosevelt to John Simpson, July 28, 1932. John Simpson Collection.

44. John Simpson to Franklin Roosevelt, September 5, 1932. John Simpson Collection.

45. Conference of Farmers Union State Officials with Governor Franklin D. Roosevelt, September 29, 1932. John Simpson Collection.

46. A. Berle to John Simpson, October 4, 1932. John Simpson Collection.

47. John Simpson to A. Berle, October 19, 1932. John Simpson Collection, 48. James A. Farley to John Simpson, October 17, 1921. John Simpson Collection.

49. Franklin Roosevelt to John Simpson, October 29, 1932. John Simpson Collection.

50. John Simpson to Franklin Roosevelt, December 17, 1932. John Simpson Collection.

51. John Simpson to Walter Leggett, November 8, 1932. John Simpson Collection.

52. John Simpson to C.H. Hyde, December 2, 1932. John Simpson Collection.

53. Guernsey Cross to C.H. Hyde, December 3, 1932. John Simpson Collection.

54. Elmer Thomas to J.H. Stsrain, December 10, 1932. Elmer Thomas Collection.

55. Edward O'Neal to C.H. Hyde, November 25, 1932. John Simpson Collection.

56. Richard Kirkendall, *Social Scientists and Farm Politics in the Age of Roosevelt*, University of Missouri, 1966, p. 54.

57. Francis G. Cutler to C.A. Hyde, (n.d.). John Simpson Collection.

58. G.W. Bohannen to Sam B. Hill, December 7, 1932. John Simpson Collection.

59. John Simpson to G.W. Bohannen, December 16, 1932. John Simpson Collection.

60. John Simpson to E.H. Everson, December 26, 1932. John Simpson Collection.

61. C.H. Hyde to Bernard Baruch, December 29, 1932. John Simpson Collection.

62. J. Edward Anderson to Franklin Roosevelt, January 20, 1932. John Simpson Collection.

63. W.B. Derk to Franklin Roosevelt, February 2, 1932. John Simpson Collection.

64. Donald Worster, *The Dust Bowl, the Southern Plains in the 1930s*, Oxford University Press, 1979, p. 13.

65. Lord, *The Wallaces of Iowa*, pp. 323–325.

66. John Simpson to Ellisson Smith, March 21, 1933. Elmer Thomas Collection.

67. Franklin Roosevelt to John Simpson, May 20, 1933. John Simpson Collection.

68. Simpson to Smith.

69. U.S. Senate, Agriculture Emergency Act to Increase Farm Purchasing Power, Hearings Before the Committee on Agriculture and Forestry, 73rd Congress, March 17–28, 1933, p. 104.

70. *Ibid.*, pp. 104–105.

71. James Wrigley to W.W. Hastings, April 15, 1933. Elmer Thomas Collection.

72. John Simpson to Franklin Roosevelt, May 6, 1933. John Simpson Collection.

73. John Simpson to Thomas Cashman, May 13, 1933. John Simpson Collection.

74. John Simpson to Franklin Roosevelt, July 10, 1933. John Simpson Collection.

75. John Simpson to Franklin Roosevelt, August 16, 1933. John Simpson Collection.

76. John Simpson to Franklin Roosevelt, September 14, 1933. John Simpson Collection.

77. Louis Howe to John Simpson, September 29, 1933. John Simpson Collection.

78. Ernest Lundeen to John Simpson, September 29, 1933. John Simpson to Ernest Lundeen, October 10, 1933. John Simpson Collection.

79. John Simpson to Franklin Roosevelt, October 24, 1933. Franklin D. Roosevelt Library.

80. Franklin Roosevelt to John Simpson, November 3, 1933. Franklin D. Roosevelt Library.

81. John Simpson to Franklin Roosevelt, November 11, 1933. Franklin D. Roosevelt Library.

82. Chicago Monitor Office to John Simpson, November 6, 1933. John Simpson to Chicago Monitor Office, November 6, 1933. John Simpson Collection.

83. Report of John A. Simpson, President, National Farmers Union, November 21, 1933. John Simpson Collection.

84. Worster, *The Dust Bowl*, p. 13.

85. Franklin Roosevelt to John Simpson, November 15, 1933. John Simpson Collection and Franklin D. Roosevelt Library.

86. John Simpson to Franklin Roosevelt, November 28, 1933. John Simpson Collection.

87. "Simpson Died as He Wished…Working," *Oklahoma Union Farmer*, April 1, 1934. Elmer Thomas Collection.

88. *Oklahoma Union Farmer*, April 1, 1934, p. 2; Elmer Thomas Collection. D.N. Kelly to Mrs. John A. Simpson, William Langer to Mrs. John Simpson, Mrs. Morris Self to Mrs. John A. Simpson, H. Wallace to Mrs. John A. Simpson, M. F. Dickinson to Mrs. John A. Simpson; John Simpson Collection.

89. Hurt, *The Dust Bowl*, pp. 89–90.

Chapter 3

1. "Months of Unrest Exploded Into 'Cow War' of '31," *Des Moines Register*, May 5, 1946, p. 5L. Interview with John Bosch, 1962. Tape in the University of Iowa Archives, p. 1–2.

2. John L. Shover, *Cornbelt Rebellion: The Farmers Holiday Movement*, Urbana: University of Illinois Press, 1965, p. 28. Frank D. Dileva, "Iowa Farm Price Revolt," *Annals of Iowa*, Vol. 32 (January 1954): 172.

3. Author interview with Virgil Johnson and Nola Eskelson, children of

A.J. Johnson, June 7, 1991. Transcript in Special Collections, Iowa State University Library.

4. Shover, *Cornbelt Rebellion*, p. 16.

5. Iowa State College, *A Century of Farming in Iowa, 1846–1946*, Ames: Iowa State University Press, 1946, p. 12.

6. U.S. Department of Commerce, *United States Census of Agriculture: 1935*, Vol. I, 1936, p. 236.

7. Milo Reno, "For Miss Prescott," Reno Papers, University of Iowa Archives, p. 1; John Shover, *Cornbelt Rebellion*, p. 25.

8. Reno, "For Miss Prescott," p. 1.

9. Lowell K. Dyson, *Red Harvest, the Communist Party & American Farmers*, Lincoln, Nebraska: University of Nebraska Press, 1982, pp. 71–72. Michael W. Flamm, "The National Farmers Union and the Evolution of Agrarian Liberalism, 1937–1946," *Agricultural History*, Summer, 1994, pp. 54–80.

10. *Ibid.*, p. 2.

11. Kenneth Finegold and Theda Skocpol, *State and Party in America's New Deal*, University of Wisconsin Press, 1955, p. 16.

12. William Hirth, "14th Annual Convention of M.F.A.," *The Missouri Farmer*, September 1, 1930, pp. 1,9.

13. Interview with John Bosch, 1962. Tape in the University of Iowa Archives, pp. 1–2.

14. Lowell Dyson, *Red Harvest, the Communist Party and American Farmers*, p. 73.

15. Bosch Interview, p. 2.

16. Dale Kramer, *The Wild Jackasses, the American Farmer in Revolt*, New York: Hastings House, pp. 208–213.

17. *Des Moines Tribune*, March 10, 1931.

18. Kramer, *The Wild Jackasses*, p. 210.

19. *Ibid.*, pp. 211–213.

20. *Ibid.*, p. 217.

21. Dan Turner, Letter to John Shover, October 15, 1961. University of Iowa Archives, p. 4.

22. "Months of Unrest Exploded into 'Cow War' of 31," *Des Moines Register*, May 5, 1946, p. 4L. Kramer, *The Wild Jackasses*, p. 218.

23. Kramer, *The Wild Jackasses*, p. 218.

24. "Months of Unrest Exploded Into 'Cow War' of '31."

25. *Ibid*; Kramer, *The Wild Jackasses*, pp. 218–219.

26. Interview with Edward E. Kennedy by Jean Choate, June 10, 1989, Iowa State University Archives, pp. 10, 11.

27. Bosch to E.H. Hillman, October 28, 1937, in Bosch Papers, cited in David Nass, editor, *Holiday: Minnesotans Remember the Farmers Holiday Association*, Marshall, Minnesota: Southwest State University Plains Press, 1984, p. xxviii.

28. Bosch Interview, April 1, 1962, p. 13.

29. Meridel LeSueur, *Crusaders: The Radical Legacy of Marian and Arthur Leseur*, St. Paul, Minnesota: Minnesota Historical Society Press, p. xxvi, cited in Linda Ford, *Women on Holiday: Gender and Midwest Agrarian Activism in the '30s*.

30. Kramer, *The Wild Jackasses*, p. 221.

31. Dyson, *Red Harvest*, p. 73. Shover, *Cornbelt Rebellion*, p. 37.

32. Theodore Saloutos and John D. Hicks, *Twentieth Century Populism*, Lincoln: University of Nebraska Press, 1951, p. 442.

33. Bosch Interview, p. 2.

34. Dyson, *Red Harvest*, p. 73.

35. "Farmers' Holiday," *Iowa Union Farmer*, March 9, 1932. Cited in Donald Lee Dougherty, "The Evolution of a Progressive: Daniel Webster Turner of Iowa," Thesis, Drake University, 1973, pp. 109–110. Shover, *Cornbelt Rebellion*, pp. 36–37.

36. Kramer, *The Wild Jackasses*, p. 224.

37. Bosch Interview, p. 3.

38. John L. Shover, "The Farmers' Holiday Strike, August 1932," *Hitting Home*, Ivan R. Dee, 1989, p. 152.

39. Des Moines, *Sunday Register*, September 23, 1931, pp. C:1, 4.

40. Bosch Interview, p. 3.

41. Bosch Interview, p. 3.

42. Kramer, *The Wild Jackasses*, pp. 225–226.

43. Frank D. Dileva, "Iowa Farm Price Revolt," *Annals of Iowa*, Vol. 32, January 1954, pp. 171–172. Dougherty, "The Evolution of a Progressive," p. 111.

44. "Clay, Union and Lincoln County Men in Favor of Present Iowa Strike," *Aberdeen Evening News*, Aug. 19, 1932, pp. 1, 8.

45. Kramer, *The Wild Jackasses*, p. 227.

46. "Boycotts, Road Picketing Used in Iowa Strike," *Aberdeen Evening News*, August 15, 1932, p. 8.

47. Turner, Letter to John Shover, Oct 15, 1961. Turner Papers, University of Iowa Archives, p. 2.

48. "Boycotts, Road Picketing Used in Iowa Strike," *Aberdeen Evening News*, p. 8.

49. *Des Moines Register*, August 15, 1932, p. 1.

50. Turner, Letter to John Shover, pp. 9–10.

51. Turner, Letter to John Shover, p. 1.

52. Kramer, *The Wild Jackasses*, p. 227. "Months of Unrest Exploded Into 'Cow War' of '31," p. 51.

53. *Ibid.*, p. 228.

54. *Sioux City Journal*, August 16, 1932, p. 1; *Minneapolis Journal*, August 16, 1932, p. 1; August 18, 1932, p. 1. Cited in Ford, *Women on Holiday*, p. 5.

55. "Four Men Injured Last Night as Picketers-Officials Clash," *Aberdeen Evening News*, August 25, 1932, p. 1; "Police Injured in Farm Holiday War," *Daily Capital Journal*, p. 1.

56. Dyson, *Red Harvest*, p. 76.

57. Kramer, *The Wild Jackasses*, p. 228–229.

58. Turner, Letter to John Shover, p. 8.

59. Kramer, *The Wild Jackasses*, pp. 229–230. "Picket Lines in Iowa Blockading Flow of Produce," *Aberdeen Evening News*, August 26, 1932, p. 1; Dyson, *Red Harvest*, p. 76.

60. *Ibid.*, p. 230; Dyson, *Red Harvest*, p. 76.

61. Turner, Letter to John Shover, p. 2.

62. "Eleven Officers Are Hurt in Clash with Picketers at Cushing, Iowa, Last Night," *Aberdeen Evening News*, August 30, 1932, p. 1.

63. "James, Iowa, Is Locality of New Strike Troubles," *Aberdeen Evening News*, September 7, 1932, p. 1.

64. Kramer, *The Wild Jackasses*, pp. 230–231.

65. "Davison County Delegation Asks Governor's Help," *Aberdeen Evening News*, August 25, 1932, p. 1.

66. "Governor's Farm Conference to Be Non-Political, Idea of Gov. Green," *Aberdeen Evening News*, September 3, 1932, pp. 1, 3.

67. "Would Hold Up Strike Until After Governors' Conference, Sept. 9," *Aberdeen Evening News*, September 3, 1932, pp. 1, 3. "Holiday Organization Is Planned," *Daily Capital Journal*, September 6, 1932, p. 1.

68. Dyson, *Red Harvest*, p. 77.

69. "Foster Fails to Talk Here, 500 Hear Woman Ally of Communist," *Des Moines Register*, September 8, 1932. Turner Collection.

70. *Ibid.*, p. 231.

71. Kramer, *The Wild Jackasses*, p. 231.

72. "Action to Aid Price of Farm Produce Asked," *Aberdeen Evening News*, September 10, 1932, pp. 1,2.

73. *Sioux City Journal*, Sept. 11, 1932, p. 1.

74. "Session Appears Near Break with Holiday Leaders," *Aberdeen Evening News*, September 11, 1932, p. 1.

75. *Ibid.*

76. Turner, Letter to John Shover, p. 4.

77. *Ibid.*

78. Daniel W. Turner, Interview with George Mills, cited in Doherty, "The Evolution of a Progressive," p. 119.

79. *Ibid.*

80. "The Farmers Strike," *The Missouri Farmer*, September 1, 1932, p. 8.

81. *Ibid.*

82. Dougherty, "The Evolution of a Progressive: Daniel Webster Turner of Iowa," p. 113.

83. "Violence Not to Be Tolerated in Newest Offense," *Aberdeen Evening News*, September 21, 1932, p. 1. "Results of New Price War Thus Far Very Quiet," *Aberdeen Evening News*, September 22, 1932, p. 1.

84. "Farm Holiday Is Watched Closely for Real Results," *Daily Capital Journal*, September 22, 1932, p. 1.

85. "Stock, Produce Exchanges Still on Normal Basis," *Aberdeen Evening News*, September 27, 1932, p. 1.

86. "Woman Thwarts Pickets—Hammer Used with Effect," *Daily Capital Journal*, October 12, 1932, p. 2.

87. "Support Asked in Closing Assembly of Farmers Union," *Aberdeen Evening News*, October 23, 1932, p. 2.

88. "Instructions for Organizing Farmers' Holiday Association...," p. 1. Dyson, *Red Harvest*, pp. 100–101.

89. Bosch Interview.

90. Harry Lux interview by John Shover, March 1, 1961. University of Iowa Archives.

91. John E. Miller, "Restrained, Respectable Radicals: The South Dakota Farm Holiday," *Agricultural History*, July, 1985, pp. 439–441.

92. *Ibid.*, p. 441.

93. Turner, Letter to John Shover, p. 7.

94. Milo Reno to W.T. Davis, January 20, 1933. Reno Collection, University of Iowa Archives, p. 1.

95. Interview with John Bosch.

96. Milo Reno, "These Are the Times That Try Men's Souls," *Farm Holiday*, Feb. 4, 1933, pp. 1–3.

97. "Editorial Comments by Milo Reno," February 27, 1933. Reno Papers, p. 1. "Roosevelt Seventh President to Come to Iowa to Select Agriculture Secretary," *Des Moines Register*, February 27, 1933, p. 2. "Henry Wallace to the Cabinet," *Des Moines Register*, February 27, 1933, p. 4.

98. Emil Loriks to Milo Reno, April 21, 1933. Reno Papers.

99. G.M. Gillette to Milo Reno, April 18, 1933.

100. Lloyd Thurston to Milo Reno, April 17, 1933. Reno Papers.

101. Frank D. DiLeva, "Attempt to Hang an Iowa Judge," *Annals of Iowa*, July 1954, pp. 337–341.

102. *Des Moines Register*, April 29, 1933, p. 1.

103. *Des Moines Tribune*, April 29, 1933, p. 1.

104. DiLeva, "Attempt to Hang an Iowa Judge," pp. 348–349.

105. Secretary to Mr. Reno to Mr. Elmer Swenson, April 26, 1933. Reno Papers.

106. Frank Eliason to Milo Reno, April 22, 1933. Reno Papers.

107. Mrs. Algot Johnson to Milo Reno, April 20, 1933. Reno Papers.

108. Milo Reno to Mrs. Algot Johnson, April 23, 1933. Reno Papers.

109. Elmer Swenson to Milo Reno, April 24, 1933. Reno Papers.

110. Milo Reno to Fred Dean, April 16, 1933. Reno Papers.

111. Milo Reno to All Officers and Members, May 8, 1933; Milo Reno to William Hirth, May 11, 1933. Reno Papers.

112. Milo Reno to Henry A. Wallace, May 9, 1933. Reno Papers.

113. Milo Reno to Thomas Cashman, May 11, 1933. Reno Papers.

114. Milo Reno to Franklin Roosevelt, May 11, 1933. Reno Papers.

115. Interview with John Bosch.

116. *Ibid.*

117. Milo Reno to Attached List of Names, May 12, 1933. Reno Papers. Dyson, *Red Harvest*, p. 111.

118. *Ibid.*

119. Milo Reno to Milan Jost, June 29, 1933. Reno Papers.

120. Milo Reno to A. P. Whitney, June 11, 1933. Reno Papers.

121. Milo Reno, Editorial, *Iowa Union Farmer*, p. 4. Reno Papers.

122. *Ibid.*

123. Milo Reno to F.M. Breed, June 11, 1933. Reno Papers.

124. Mrs. Charles Wedholm to Milo Reno, July 5, 1933. Reno Papers.

125. Milo Reno to John A. Simpson, June 17, 1933. Reno Papers.

126. Charles Ray, Jr., to Milo Reno, July 22, 1933. Reno Papers.

127. John Bosch to Milo Reno, July 18, 1933. Reno Papers.

128. Milo Reno to John Bosch, August 17, 1933. Reno Papers.

129. Milo Reno to John Bosch, August 22, 1933. Reno Papers.

130. Milo Reno, "The Future of the National Farmers' Holiday Association," *Iowa Union Farmer*, July 24, 1933. Reno Papers, p. 1.

131. F.W. Murphy to Milo Reno, August 13, 1933. Reno Papers.

132. Milo Reno to C.H. Larison, Estherville, Iowa, August 17, 1933; W.C. Condon to Milo Reno, August 31, 1933; Milo Reno to E.E. Kennedy, August 22, 1933. Reno Papers.

133. Milo Reno to President Roosevelt, September 1, 1933. Reno Papers.

134. Milo Reno to Richard Bosch, September 7, 1933. Reno Papers.

135. Milo Reno, Editorial, *Iowa Union Farmer*. Reno Papers, pp. 1–6.

136. Milo Reno, Editorial, *Farm Holiday News*, pp. 1–3 Reno Papers.

137. M.L. Glarum to Milo Reno, September 11, 1933. Reno Papers.

138. *Ibid.*

139. Mrs. Chris Linnertz to Jess Sickler, September 15, 1933. Reno Papers.

140. Fred Schmidt to Milo Reno, September 21, 1933. Reno Papers.

141. Khaki Shirts of America to Milo Reno, September 11, 1933; Milo Reno to Art Smith, Commander, Khaki Shirts of America, September 15, 1933. Reno Papers.

142. Milo Reno to W.C. Condon, September 16, 1933. Reno Papers.

143. Milo Reno to Olson, September 18, 1933. Reno Papers.

144. "Threat of Farm Strike at Meet," *Aberdeen Evening News*, September 21, 1933, p. 1. "NRA Blue Eagle Code Demanded by Agriculture," *Aberdeen Evening News*, September 23, 1933, p. 1.

145. Milo Reno to Franklin D. Roosevelt, September 23, 1933. Reno Papers.

146. Milo Reno to Usher Burdick, September 27, 1933; Milo Reno to Mrs. Chris Linnertz, October 3, 1933. Reno Papers.

147. M.H. McIntyre to Milo Reno, September 25, 1933. Reno Papers.

148. Bosch Interview. Milo Reno to Mrs. Chris Linnertz, October 3, 1933. Reno Papers.

149. John A. Simpson to Milo Reno, October 4, 1933. Reno Papers.

150. Milo Reno to George M. Griffin, October 6, 1933. Reno Papers.

151. Milo Reno to John A. Simpson, October 6, 1933. Reno Papers.

152. Dyson, *Red Harvest*, p. 116.

153. E.E. Kennedy, John H. Bosch and Harry C. Parmenter to Milo Reno, October 10, 1933. Reno Papers. John Bosch remembers that in their meeting with Roosevelt, John Simpson got to talking on the quantity theory of money and thus the discussion did not focus as much on the agricultural code as it might have, although Bosch conceded that he did not know if it really made any difference. "I don't think that we would have gotten any place. It was just something they were not going to accept, that's all." Bosch Interview.

154. Milo Reno to Clyde Herring, October 10, 1933. Reno Papers.

155. Milo Reno to the State Presidents and Secretaries of the National Farmers Holiday Association, October 10, 1933. Reno Papers.

156. *Ibid.*

157. Bert Salisbury to Milo Reno, October 11, 1933. Reno Papers.

158. W.R. Hogan to Milo Reno, October 12, 1933. Reno Papers.

159. Charlie Peters to Milo Reno, October 14, 1933. Reno Papers.

160. Wm. Langer to Milo Reno, October 17, 1933; Milo Reno to William Langer, October 17, 1933; Usher Burdick to Milo Reno, October 17, 1933; Milo Reno to Olaf Braatelien, October 16, 1933. Reno Papers.

161. Members Plymouth County Farm Holiday to Milo Reno, October 16, 1933. Reno Papers.

162. Henry Weber to Milo Reno, October 16, 1933. Reno Papers.

163. E.T. Nahshine to Milo Reno, October 17, 1933. Reno Papers.

164. I.B. Rabinold to Milo Reno, October 18, 1933. Reno Papers.

165. Milo Reno to Mrs. W.A. Jasperson, October 18, 1933. Reno Papers.

166. Tom Berry to Milo Reno, October 19, 1933; Charles W. Bryan to Milo Reno, October 20, 1933; Milo Reno to Charles Bryan, October 20, 1933. Reno Papers.

167. George A. Shaffer to Milo Reno, October 20, 1933. Reno Papers.

168. Emil Loriks to Milo Reno, October 21, 1933. Reno Papers.

169. D.D. Collins to Milo Reno, October 20, 1933. Reno Papers.

170. E.E. Kennedy to Milo Reno, October 21, 1933. Reno Papers.

171. *Ibid.*, p. 2.

172. Milo Reno to State Presidents of the Farmers Holiday Association, October 21, 1933. Reno Papers, p. 1.

173. *Ibid.*

174. Milo Reno to John Erp, October 25, 1933. Reno Papers. He also gave an address over *NBC Farm and Home Hour* on October 26. A. R. Williamson to Milo Reno, October 25, 1993. Reno Papers.

175. Hans Sorby to Milo Reno, October 23, 1933; Milo Reno to Hans Sorby, October 27, 1933. Reno Papers.

176. Milo Reno to J.O. Thompson, October 27, 1933. Reno Papers.

177. Orle Oian to Mil Reno, October 30, 1933. Reno Papers.

178. R.A. Smalley to Milo Reno, October 30, 1933. Reno Papers.

179. Ernest Volz to Milo Reno, October 30, 1933. Reno Papers.

180. James Montgomery to Milo Reno, October 31, 1933. Reno Papers.

181. Milo Reno to Fred Fester, November 9, 1933. Reno Papers.

182. Milo Reno, Editorial, *Farm Holiday News*. Reno Papers.

183. Milo Reno to George A. Tool, November 6, 1933. Reno Papers.

184. Milo Reno to R. S. Nortman, November 9, 1933. Reno Papers.

185. Milo Reno to Mrs. C. Frank Powell, November 9, 1933. Reno Papers.

186. Milo Reno, Editorial, *Iowa Union Farmer*, November 13, 1933. Reno Papers.

187. Richard Bosch to Milo Reno, November 4, 1933; Milo Reno to Richard Bosch, November 6, 1933. Reno Papers.

188. Fred Plueger to Milo Reno, November 25, 1933. Reno Papers.

189. *Ibid.*

190. Henry C. Blome to Milo Reno, November 25, 1933; Henry C. Blome to Milo Reno, November 27, 1933. Reno Papers.

Chapter 4

1. Clifford R. Hope, Sr., "Kansas in the 1930s," *Kansas Historical Quarterly*, 36 (Spring 1970), p. 11.

2. James C. Carey, "William Allen White and Dan D. Casement on Government Regulation," *Agricultural History*, Vol. 33 (January 1959), p. 16. "Captain Dan D. Casement," n.d. Casement Papers, University Archives, Kansas State University.

3. Casement, "A Farmer on the Farm Problem," *The Round Up*, Oak Park, Illinois (14 March 1939), p. 6.

4. Casement to Donald Van Vleet, 26 April 1936. National Farmers Processing Tax Recovery Association Records (NFPTRA Records), Special Collections, Parks Library, Iowa State University.

5. *Ibid.*

6. "Links Farm Group to Liberty League," *The New York Times*, 11 April 1936, p. 6 L.

7. James C. Carey, "The Farmers' Independence Council of America, 1935–1938," *Agricultural History*, Vol. 35 (1961) p. 70.

8. "Organization and Declaration of Principles," p. 1–2. Casement Papers.

9. Stanley Morse to Dan Casement, 15 May 1935. Casement Papers.

10. *Ibid.*

11. *Ibid.*

12. *Ibid.*

13. Morse to Casement, 21 June 1935. Casement Papers.

14. Morse to Casement, Wilcox, Crawford and Dorsett, 22 June 1935. Casement Papers.

15. Morse to Casement, 23 June 1935. Casement Papers.

16. Morse to Casement, 6 July 1935. Casement Papers.

17. Morse to Casement, 8 July 1935. Casement Papers.

18. Edward N. Wentworth to Casement, 11 July 1935. Casement Papers.

19. Wentworth to Casement, 12 July 1935. Casement Papers.

20. Morse to Casement, 24 September 1935. Casement Papers.

21. Morse to Casement, 15 July 1935. Casement Papers.

22. Morse to Casement, 27 August 1935. Casement Papers.

23. Morse to Casement, 31 August 1936. Casement Papers.

24. Morse to Casement, 10 October 1936. Casement Papers.

25. H. Alexander Smith to Casement, 21 October 1935. Casement Papers.

26. Morse to Casement, 23 December 1935; Lammot DuPont to Morse, 20 December 1935. Casement Papers.

27. S.D. Townsend to Morse, 21 December 1935; J. Thompson Brown to Morse, 20 December 1935. Casement Papers.

28. Morse to Casement, 23 June 1935. Casement Papers.

29. Casement to Wm. Whitfield Woods, 22 July 1935. Casement Papers.

30. Morse to Casement, 24 July 1935. Casement Papers.

31. Morse to Casement, 31 August 1935. Casement Papers.

32. Morse to Casement, 10 October 1935. Casement Papers.

33. Morse to member of Farmers Independence Council, 1 April 1936. Casement Papers.

34. Morse to Casement, 21 September 1935. Casement Papers.

35. Farmers Independence Council to Henry A. Wallace, 25 September 1936. Casement Papers.

36. *Ibid.*

37. "What the President Said at Fremont and How His Statements Line Up..." n.d. Casement Papers.

38. Morse to Casement, 5 August 1935. Casement Papers.

39. Theo. H. Lampe to Casement, 16 July 1935. Casement Papers.

40. Theo. Lampe to Casement, 1 August 1935. Casement Papers.

41. Lawrence B. Evans, *Cases on American Constitutional Law*, Chicago: Callaghan & Company, 1952, p. 140.

42. Farmers Independence Council to Members and Friends, "Our Job," 9 January 1936. Casement Papers.

43. Chester Davis to Casement, 8 January 1936. Casement Papers.

44. Casement to Franklin D. Roosevelt, 11 January 1936. Casement Papers.

45. Smith to Casement, 13 January 1936. Casement Papers.

46. Casement to Henry A. Wallace, 17 March 1936. Casement Papers.

47. Casement to Hoover, 31 March 1936. Hoover Papers.

48. Kansas Farm Bureau, 3 May 1935. Casement Papers.

49. Alden A. Potter to Casement, 17 May 1935. Casement Papers.

50. Mrs. C.M. Buchanan to Casement, 5 May 1935. Casement Papers.

51. Morse to Casement, 26 June 1935. Casement Papers.

52. Morse to Casement, "Memorandum in Re National Farm Program," 12 November 1935. Casement Papers.

53. Smith to Casement, 13 January 1936. Casement Papers.

54. W.M. Jardine to Kurt Grunwald, 6 January 1936. Casement Papers.

55. Smith to Hoover, 21 October 1935. Hoover Papers.

56. Hoover to Morse, 27 January 1936. Casement Papers.

57. Morse to Hoover, 10 February 1936. Hoover Papers.

58. Charles W. Burkett to Morse, 10 November 1935. Casement Papers.

59. Morse to Casement, 16 January 1936. Casement Papers.

60. Morse to Burkett, 18 January 1936. Casement Papers.

61. Morse to Casement, 4 February 1936. Casement Papers.

62. Morse to Casement, 10 February 1936. Casement Papers.

63. Morse to the Directors, Farmers Independence Council of America, 10 February 1936. Casement Papers.

64. Morse to Casement, 7 April 1936. Casement Papers.

65. Morse to Hoover, 9 April 1936. Hoover Papers; Hearings Before a Special Committee to Investigate Lobbying Activities, U.S. Senate, 74th Congress, 1st Session, Part 5, March and April 1936, pp. 1892–1893.

66. Hoover to Morse, 15 April 1936. Hoover Papers.

67. Casement to Hugo Black, 15 April 1936. Casement Papers.

68. Hearings, pp. 1881–1889.

69. Morse to Members of the Farmers' Independence Council of America, 20 April 1936. Casement Papers.

70. Morse to Members of the Farmers' Independence Council, 20 April 1936. Hoover Papers.

71. Hearings, p. 2042.

72. Morse to Members of the Farmers' Independence Council, 20 April 1936.

73. "Links Farm Group to Liberty League," New York Times, 11 April 1936, p. 6L.

74. Morse to Hoover, 9 April 1936. Hoover Papers; Hoover to Morse, 18 April 1936. Hoover Papers.

75. Casement to Alf. M. Landon, 12 June 1936. Casement Papers.

76. Morse "Memorandum," 27 June 1936. Casement Papers.

77. Morse to Casement, 29 June 1936. Casement Papers.

78. Morse to Casement, 2 July 1936. Casement Papers.

79. Morse to Harrison E. Spangler, 2 July 1936. Hoover Papers.

80. Hoover to Morse, 9 July 1936. Hoover Papers.

81. Morse to Hoover, 18 July 1936. Hoover Papers.

82. Morse to Jasper Crane, 2 July 1936. Casement Papers.

83. Morse to Arthur Curtis, 17 July 1936. Hoover Papers.

84. Hoover to Morse, 20 July 1936. Hoover Papers.

85. Kurt Grunwald to Casement, 6 August 1936. Casement Papers.

86. Morse to Hoover, 20 August 1936. Hoover Papers; Casement, 6 October 1936. Casement Papers.

87. Hoover to Morse, 25 August 1936. Hoover Papers.

88. "Wearing False Whiskers," New York Times, 4 October 1936, p. 42.

89. Morse to Casement, 11 November 1936. Casement Papers.

90. Morse to Casement, 11 November 1936. Casement Papers.

91. Morse to Casement, 9 February 1937. Morse, "Preliminary Financial Statement," 9 February 1936. Casement Papers.

92. Morse to Casement, 26 February 1937; Morse to Casement, 6 May 1937. Casement Papers. "Call for Action, the Supreme Court and Constitution of the United States UNDER ATTACK," n.d.; Morse to Hoover, 9 March 1937; Hoover to Morse, 12 March 1937; Morse to Hoover, 22 March 1937. Hoover Papers.

93. Editor, Saturday Evening Post, to Casement, 17 September 1937; Merle Thorpe, Nation's Business, to Casement, 12 October 1937; David Page, The Digest, to Casement, 21 October 1937. Casement Papers.

94. Casement to Kent, 6 June 1937; Smith to Casement, 7 June 1937; Alison Reppy to Casement, 7 June 1937; Smith to Casement, 26 June 1937. Casement Papers.

95. Morse to Casement, 11 December 1937. Casement Papers. "Senate Farm Group Drafts Bill for Five-Crop Control; Defies Roosevelt on Cost," New York Times, November 22, 1937, p. 1.

96. Ibid; "Grange Gives Farm Policy," Detroit News, 24 November 1939, p. 12.

97. Hawkinson to Casement, 18 February 1938. Casement Papers.

98. Morse to Casement, 14 February 1938. Casement Papers.

Chapter 5

1. Joseph Frazier Wall, "The Iowa Farmer in Crisis, 1920–1936," *Annals of Iowa*, Vol. 47 (Summer 1983), pp. 118–127.

2. Hattie Kroger to Donald Van Vleet, March 19, 1936. National Farmers Process Tax Recovery Association Records (NFPTRA), Archives of Iowa State University.

3. Theodore Saloutos, *The American Farmer and the New Deal*, Ames: Iowa State University, 1982, p. 55. According to the terms of the act, seven major commodities were subject to control: cotton, corn/hogs, wheat, dairy products, tobacco and rice.

4. C. P. Rusch to Secretary Henry A. Wallace, November 27, 1933, NARG 145, AAA 1933–1935. Subject Correspondence, Hogs P. T.

5. John Marnach to Secretary Henry A. Wallace, Dec. 7, 1933, NARG 145, AAA 1933–1935. Subject Correspondence, Hogs P. T.

6. Walter Bartels to Secretary Henry Wallace, January 13, 1934, NARG 145, AAA 1933–1935. Subject Correspondence, Hogs P. T.

7. John L. Shover, *Cornbelt Rebellion: The Farmers Holiday Movement*, Urbana: University of Illinois Press, 1965, pp. 142–145. William P. Tucker, "Populism Up-to-Date: The Story of the Farmers' Union," *Agricultural History*, Vol. 21 (Oct. 1947), p. 199.

8. Gladys Baker, *The County Agent*, Chicago: University of Chicago Press, 1939, p. 15.

9. White, *Milo Reno*, pp. 67–69, 181.

10. Schuyler, "The Hair-Splitters," p. 408.

11. "Huey Long Comments on Resolution to Oust Wallace," *Unionist and Public Forum*, May 17, 1934, pp. 1, 4.

12. Radio Address by Milo Reno over radio station WHI, March 17, 1935. NFPTRA Records.

13. C.M. Perrin, "The Paradise of Dreams vs. the Stern Realities of Life or Myths vs. Facts," speech given at Monona County Farmers Union Meeting, Feb. 7, 1934. A.J. Johnson Scrapbook.

14. Edward E. Kennedy, Farmers Washington Service, October 25, 1937. NFPTRA Records.

15. Reno to Erchaker, January 13, 1936. Reno Papers.

16. "Minutes of Iowa Farmers Union Councilors' Meeting," June 24, 1942, p. 7. Reno Papers.

17. "Minutes of the Meeting," March 12, 1936; A.J. Johnson to Milo Reno, March 13, 1936. NFPTRA Records.

18. "Minutes of the Meeting," March 12, 1936. NFPTRA Records.

19. Interview with Virgil Johnson and Nola Eskelson, children of A.J. Johnson, June 7, 1991. Transcript on file in Special Collections of Iowa State University.

20. "Minutes of Meeting," March 12, 1935. NFPTRA Records.

21. *Ibid.*

22. Charter Members of Iowa Division Farmers Process Tax Recovery Association, NFPTRA Records.

23. "Monona F-L, to Meet at Onawa May 12"; " F.L. Federation Meeting Last Week," *Unionist and Public Forum*, May 12, 1936, p. 2.

24. A.J. Johnson to Milo Reno, March 13, 1936; G.L. Harrison, Christian Grell and Henry Preyt to Milo Reno, March 12, 1936. Reno Papers.

25. Hattie Kroeger to Donald Van Vleet, June 21, 1936. NFPTRA Records.

26. O.D. Ferry to Donald Van Vleet, June 21, 1936. NFPTRA Records.

27. "States Contacted," NFPTRA Records.

28. Statement of A.J. Johnson, Secretary of the NFPTRA, before the House Committee on Agriculture, NFPTRA Records.

29. Donald Van Vleet to George C. Stokes, June 16, 1936. NFPTRA Records.

30. "Minutes of Iowa Farmers Union Councilors' Meeting," June 24, 1941, p. 7, Reno Papers. Donald Van Vleet to Andrew Jensen, January 13, 1938, NFPTRA Records. (Although Shover, in *Cornbelt Rebellion*, pp. 171–172, indicates that Van Vleet received money from Dan Casement, president of the Liberty League, my reading of the source material suggests that Van Vleet requested money from Casement. Other letters in the NFPTRA Records show that Van Vleet hoped to get money but was disappointed.)

31. F.M. Simpson to Donald Van Vleet, December 10, 1937; Donald Van Vleet to F.M. Simpson, Dec. 14, 1937. NFPTRA Records.

32. Donald Van Vleet to John Erp, July 15, 1937. NFPTRA Records.

33. George De Bar to Donald Van Vleet, May 26, 1937. NFPTRA Records.

34. Donald Van Vleet to Frederick Free, April 28, 1937; Frederick Free to Donald Van Vleet, June 3, 1937; Minutes of Annual Meeting, Farmers Process Tax Recovery Association, April 19, 1938. NFPTRA Records.

35. Donald Van Vleet to Representative Hubert Utterback, May 1936. NFPTRA Records.

36. Dayton Tax Service, Washington, Iowa, "To the Solicitors of the Processing Tax Recovery Association," March 17, 1937; Donald Van Vleet to William Lemke, February 26, 1937; William Lemke to Donald Van Vleet, March 2, 1937; Donald Van Vleet to Dale Kramer, April 21, 1937. NFPTRA Records.

37. Donald Van Vleet to William Lemke, Feb. 26, 1937; William Lemke to Donald Van Vleet, March 2, 1937. NFPTRA Records.

38. Donald Van Vleet to Dale Kramer, April 21, 1937. NFPTRA Records.

39. Dan Casement to Donald Van Vleet, April 26, 1936. Casement Papers.

40. Donald Van Vleet to John Erp, May 10, 1937. NFPTRA Records.

41. Edward E. Kennedy to A.J. Johnson, June 23, 1937. NFPTRA Records.

42. "National Farm Leaders Attend Rites for Reno," *Daily Republic*, May 9, 1936, p. 12.

43. Tucker, "Populism Up-to-Date," p. 209.

44. Edward E. Kennedy interview, June 10, 1989; Shover, *Cornbelt Rebellion*, p. 210.

45. Donald Van Vleet to Edward E. Kennedy, July 3, 1937. NFPTRA Records.

46. Statement of A.J. Johnson before the House Committee on Agriculture, 1939. NFPTRA Records.

47. Donald Van Vleet to John Erp, August 3, 1937, NFPTRA Records.

48. Edward C. Blackorby, *Prairie Rebel, the Public Life of William Lemke*, Lincoln: University of Nebraska Press, 1963, p. 200.

49. E.E. Kennedy to Donald Van Vleet, August 4, 1937. NFPTRA Records.

50. Donald Van Vleet to Paul Bock, September 8, 1937. NFPTRA Records.

51. Wm. Lemke to E.E. Kennedy, October 22, 1937. Lemke Papers.

52. E.E. Kennedy to Wm. Lemke, Nov. 3, 1937. Lemke Papers.

53. E.E. Kennedy to Donald Van Vleet, Nov. 30, 1937. NFPTRA Records.

54. Edward E. Kennedy to the State Managers of the NFPTRA Ass'n., Nov. 30, 1937, "The Producer Paid Unconstitutional AAA Tax on Hogs Says Bureau of Agriculture Economics." NFPTRA Records.

55. *Ibid.*

56. *Ibid.*

57. *Ibid.*

58. "South Dakotans Remember," Emil Loriks interview by Paul O'Rourke, edited by Geo. Wolff and Joseph N. Cook, *South Dakota History*, Summer 1989, pp. 227–228.

59. Paul Appleby to Conrad Beisser, Dec. 23, 1936, NARG 16, Secy's Office Correspondence, 1936, Taxes.

60. Henry A. Wallace (Secretary's File Room, signed) to Postmaster General, January 14, 1937, NARG 16, Secy's Office Correspondence, 1937, Taxes.

61. K. P. Aldrich to Dept. of Agriculture, April 12, 1938, NARG 16, General Correspondence, 1938, Taxes.

62. Lee M. Gentry to Claude Wickard, October 6, 1937; Claude Wickard, "Memorandum for Mr. Lee Gentry," Oct. 25, 1937, NARG 145, National B-IT.

63. John Maier to D.B. Gurney, March 9, 1938. NFPTRA Records.

64. Andrew Hoganson to D.B. Gurney, April 13, 1938. NFPTRA Records.

65. Mastin G. White, Statement, January 11, 1938, NARG 16, General Correspondence, 1938, Taxes.

66. Paul Appleby to Sioux City Live Stock Exchange, Feb. 8, 1938; John W. Carey to Paul Appleby, Feb. 9, 1938; Paul Appleby to John W. Carey, Feb. 10, 1938; NARG 16, General Correspondence, 1938, Taxes.

67. "Warns Farmers About Refunds," *Yankton Press*, Feb. 9, 1938, p. 1; "Hog Processing Tax," *Farmers Union Herald*, March 1938, p. 6; "The Hog Processing Tax Recovery Association," *Farmers Union Herald*, April 1938, p. 5; "Statement Warns Farmers Against Misleading Solicitation for Processing Tax Refunds," *Dakota Farmer*, March 26, 1938, p. 1; "More About Hog Processing Tax Recovery Associations," *Farmers Union Herald*, May, 1938, pp. 1, 2.

68. A.I.G. Valentine to Andrew Peterson, March 10, 1938. NFPTRA Records.

69. Geo. Hasenbank to D.B. Gurney, March 8, 1938. NFPTRA Records.

70. Donald Van Vleet to J.E. Iverson, Feb. 15, 1938. NFPTRA Records.

71. Edward E. Kennedy to Robert Spencer, Feb. 22, 1938. NFPTRA Records.

72. E.E. Kennedy to the State Managers of the NFPTRA Ass'n, November 30, 1937. NFPTRA Records.

73. E.E. Kennedy to Donald Van Vleet, December 13, 1937. NFPTRA Records.

74. Wm. Lemke to Andrew Trovaton, December 10, 1937. Lemke Papers.

75. E.E. Kennedy to Donald Van Vleet, January 10, 1938, Kennedy Payments. NFPTRA Records.

76. Lemke's Speech, *Congressional Record*, January 24, 1938. NFPTRA Records.

77. E.E. Kennedy to Donald Van Vleet, February 8, 1938. NFPTRA Records.

78. C.J. Lee to Wm. Lemke, February 5, 1938; Wm. Lemke to C.J. Lee, February 9, 1938. Lemke Papers.

79. E.E. Kennedy to Donald Van Vleet, February 8, 1938. NFPTRA Records.

80. Farmers Process Tax Recovery Association Board Meeting, March 19, 1938. NFPTRA Records.

81. Donald Van Vleet to E.E. Kennedy, March 23, 1938. NFPTRA Records.

82. Mary Puncke to Donald Van Vleet, March 7, 1938. NFPTRA Records.

83. U.S. Congress, Senate, *Refund of Processing Tax on Hogs to Producers*, Hearings before the Committee on Agriculture on S.J. Resolution 202, 75th Cong., 2nd Sess., 1938, pp. 4–5.

84. *Ibid.*, p. 2.

85. *Ibid.*, p. 5.

86. *Ibid.*, p. 8; Representative Clifford Hope, "Letter to the President of Swift and Co.," *Congressional Record*, February 3, 1936. NFPTRA Records.

87. U.S., Congress, Senate, *Refund of Processing Tax*, p. 8.

88. *Ibid.*, pp. 8, 9.

89. *Ibid.*, p. 10.

90. *Ibid.*

91. *Ibid.*, p. 16.

92. *Ibid.*, p. 18.

93. *Ibid.*, pp. 20–23.

94. *Ibid.*, pp. 32–33.

95. U.S. Congress, House of Representatives, *Certain Tax Refunds,* Hearings before a Special Subcommittee of the Committee on Agriculture on S2601 and other measures, 75th Cong., 2nd sess., 1938, pp. 2–3.

96. *Ibid.*

97. *Ibid.*, p. 5.

98. *Ibid.*, p. 72.

99. *Ibid.*, pp. 92–93.

100. *Ibid.*, p. 96.

101. The author has searched the Lemke files and has not found quite that number of letters from farmers on the processing tax. There have been approximately 20 or 30 letters found on that subject. Two possibilities exist. Either many of the letters were thrown away and not put into the files, or Lemke was exaggerating.

102. U.S. Congress, House of Representatives, *Certain Tax Refunds,* pp. 98–99.

103. *Ibid.*

104. *Ibid.*, pp. 63–64.

105. *Ibid.*, p. 99.

106. E.E. Kennedy, "Washington Letter," March 24, 1938, p. 3. NFPTRA Records.

107. Donald Van Vleet and A.J. Johnson "To All State Managers and County Committee Men," March 25, 1938. NFPTRA Records.

108. Edward Borkwoldt to NFPTRA, March 30, 1938. NFPTRA Records.

109. Donald Van Vleet to E.E. Kennedy, March 31, 1938. NFPTRA Records.

110. E.E. Kennedy "To All Representatives, State and County," April 6, 1938. NFPTRA Records.

111. "Processing Tax Refunds Must Not Go to Professional Racketeers," *Appendix to the Congressional Record,* U.S. Congress, *Congressional Record,* 75th Cong., 2nd sess., 1938, p. 1487.

112. W. Skeels to Wm. Lemke, June 11, 1938; E.E. Kennedy to Wm. Lemke, June 16, 1938. Lemke Papers.

113. D.B. Gurney to Albert J. Schmidt, June 23, 1938; D.B. Gurney to F.E. Fowler, June 21, 1938. NFPTRA Records.

114. Kennedy, *The Fed and the Farmer,* p. 101.

115. Family of A.J. Johnson interview, June 7, 1991. Transcript in Special Collections, Library Iowa State University.

116. A.J. Johnson to E.A. Zupke, January 14, 1939. NFPTRA Records.

117. Robert Spencer, "Call of Committee on New Farm Organization," January 16, 1939. NFPTRA Records.

118. "H.J. Res. 138, 76th Congress, 1st Session." Lemke Papers.

119. Wm Lemke to M.C. Scheuffele, Feb. 13, 1939. Lemke Papers.

120. Wm. Lemke to Glenn T. Stebbins, Feb. 14, 1939. Lemke Papers.

121. *Congressional Record, House,* March 24, 1939, p. 3233.

122. Theodore Saloutos and Johns D. Hicks, *Twentieth Century Populism, Agricultural Discontent in the Middle West, 1900–1939,* Lincoln: University of Nebraska Press, 1951, pp. 534–535. They state that letters from Farmers Union leaders came as a "great surprise because the proponents of the measure had counted on the Farmers Union for their strongest support." However, evidence in the Lemke and Kennedy paper indicates that to them the opposition of some of the leaders of the Farmers Union must have come as no surprise.

123. U.S. Congress, Senate, Subcommittee of the Committee on Agriculture and Forestry, *Refund of Processing Tax on Hogs, S. J. Res. 66*, May 16 and 17, 1939, 76th Cong., 1st sess., 1939, pp. 6–16.

124. *Ibid.*, pp. 27–33.

125. *Ibid.*, pp. 34–38.

126. *Ibid.*, p. 40.

127. A.J. Johnson to W.E. Wright, July 11, 1939. NFPTRA Records.

128. Wm. Lemke to Ed Johnson, July 25, 1939. Lemke Papers.

129. E.E. Kennedy, "Refund of the Hog Processing Tax Never Returned," *Union Farmer*, January 1941, pp. 1,7.

130. Wm. Lemke to Arthur J. Lanz, May 1, 1940. Lemke Papers.

131. Kennedy, "Refund of the Hog Processing Tax Never Returned," p. 7.

132. Wm. Lemke to E.L. Marlowe, May 17, 1940. Lemke Papers.

133. W.O. Skeels to Wilfred Wene, June 4, 1940; W.O. Skeels to Oliver Lee, June 14, 1940. Lemke Papers.

134. Blackorby, *Prairie Rebel*, pp. 244–257.

135. U.S. Congress, Senate, *Secondary Deficiency Appropriation Bill*, 76th Cong., 2nd sess., 1940, pp. 20–33.

136. A.J. Johnson to Albin Hultgren, March 3, 1941. NFPTRA Records.

137. A.J. Johnson to Carl Timm, February 25, 1941; A.J. Johnson to Jas. W. Mehaleck, March 3, 1941. NFPTRA Records.

138. A.J. Johnson to Albin Hultgren, March 3, 1941. NFPTRA Records.

139. E.E. Kennedy to A.J. Johnson, March 4, 1941. NFPTRA Records.

140. U.S. Congress, Senate, Committee on Agriculture and Forestry, *Refund of Hog Processing Tax*, May 6, 7, 12, 1941, 77th Cong., 1st sess., 1941, pp. 2–5.

141. *Ibid.*, p. 6.

142. *Ibid.*, p. 7.

143. *Ibid.*, pp. 8–15.

144. *Ibid.*, pp. 15–20.

145. *Ibid.*, pp. 21–29.

146. Kennedy, *Kennedy's Washington Letter*, June 25, 1941. NFPTRA Records.

147. Herbert E. Gaston to H. P. Fulmer, July 25, 1941. NFPTRA Records.

148. E.E. Kennedy to Fred Gilchrist, Sept. 24, 1941. Kennedy Papers.

149. Kennedy, *The Farmer and the Fed*, pp. 129–146.

150. Fred Gilchrist to E.E. Kennedy, October 2, 1941. Kennedy Papers.

151. Fred E. Dutton to A.J. Johnson, January 27, 1942. NFPTRA Records.

152. Christian Grell to Farmers Process Tax Recovery Association (no date). Kennedy Collection.

153. Christian Grell to E.E. Kennedy, February 5, 1942. Kennedy Collection.

154. Recovery Association member to Christian Grell, February 3, 1942. NFPTRA Records.

155. Robert B. Chipperfield to George Snively, February 3, 1942. NFPTRA Records.

Chapter 6

1. South Dakota Crop and Livestock Reporting Services, *South Dakota Agricultural Statistics*, 1940 and 1941, Sioux Falls: 1945, p. 15.

2. Herbert S. Schell, *History of South Dakota*, Lincoln: University of Nebraska Press, 1961, p. 283.

3. *Ibid.*, p. 284.

4. South Dakota Crop and Livestock Reporting Service, p. 17.

5. Dale Lewis, "WNAX Rushes Aid to Stricken People," *Dakota West*, Vol. X, p. 15.

6. "Governor Charged with Negligence," *Aberdeen Evening News*, Feb. 9, 1932, p. 1.

7. Jean Choate, "The National Farmers Process Tax Recovery Association in Minnesota and Iowa," *Minnesota History*, Fall 1990, pp. 100–111.

8. Lemke's Speech, Congressional Record, January 24, 1938, National Farmers Process Tax Recovery Association (NFPTRA) Records. Department of Special Collections, Parks Library, Iowa State University, Ames. Except where otherwise noted, all correspondence cited is from the NFPTRA collection. Most of the Gurney correspondence is in boxes 16–28.

9. William Lemke to O.S. Eastvold, Volin, S.D., October 26, 1938, Orin G. Libby Historical Manuscripts Collection, University of North Dakota Library, Grand Forks, North Dakota. Collection #13, box 14, file folder 2.

10. D.B. Gurney to Bryce Sharp, Arco, Minn., January 31, 1938. NFPTRA Records.

11. Interview with Charles Ramon Gurney (Grandson of D.B. Gurney), Iowa City, August 2, 1990.

12. Matt Fellers, Bluffton, Minn., to D.B. Gurney, January 28, 1938.

13. R.H. Block of RD7 Menomonie, Wis., March 2, 1938.

14. Charles Matz, Oconto, Neb., to D.B. Gurney, April 11, 1938.

15. NFPTRA Collection. Minnesota actually had 1216 replies, followed by South Dakota at 729; Nebraska at 509; Iowa at 487; North Dakota at 207; and Wisconsin at 70.

16. Charles Ramon Gurney Interview, August 2, 1990.

17. Ernest E. Fenske, Rt. 2, Blue Earth, Minn., to D.B. Gurney.

18. C.W. Valentine, R.2, De Graff, Minn., to D.B. Gurney, Aug 8,1938.

19. F.A. Gastanczik, Kasota, Minn., to D.B. Gurney, Feb. 12, 1938.

20. The ratio of non-signers to signers varies with the dates of the letters. In the earliest letters received by Gurney there were twice as many non-signers as signers, even when the category included all those who had signed up with the AAA program for any period at all. In later letters the ratio was nearly even. Of all the farmers who indicated in their letters whether they were signers or non-signers, and many did not, there were 587 signers and 821 non-signers.

21. Joseph Shirer, Rockham, S.D., to D.B. Gurney, Feb. 2, 1938.

22. D.B. Gurney to Gilbert Gunderson, Bryant, S.D., Feb. 22, 1938.

23. D.B. Gurney to A.N. Wilbert, Beresford, S.D., March 8, 1938.

24. D.B. Gurney to Reinhold Brickman, Colome, S.D., March 26,1938.

25. D.B. Gurney to Albert Dewitz, Tappen, N.D., March 26, 1938.

26. D.B. Gurney to Henry Foreston, Foreston, Minn., March 28, 1938.

27. The amount farmers paid depended on the size of the claim. For total claims of $100 or less, a farmer would pay two dollars. If the total claim was $200 to $400, a farmer was to pay two percent of the claim. If the total claim was more than $400 and less than $900, a farmer would pay eight dollars plus 1 1/2 percent of the amount in excess of the $400. If the total claim was more than $900, the farmer would pay $15.50 plus one percent of everything in excess of $90. Preliminary Information Blank, NFPTRA collection.

28. *Ibid.*

29. Nick Hoff, Clear Lake, S.D., to D.B. Gurney, Feb. 2, 1938; Wm. Krumrei, RFD 1, Harrisburg, S.D. to D.B. Gurney, January 29, 1938.

30. Warren Bros, River Falls, Wis., to D.B. Gurney, Feb. 17, 1938.

31. Michael Traxinger, Herreid, S.D., to D.B. Gurney, April 1, 1938.

32. Frank N. Kinney, Gary, S.D., to D.B. Gurney, Feb. 9, 1938.

33. Max Fiegen, Dell Rapids, S.D., to D.B. Gurney, Feb. 3, 1938. NFPTRA Collection.

34. E.A. Hoegh, Hampton, Neb., to D.B. Gurney, March 29, 1938.

35. C.C. Ogborn, Office Manager of John Morell & Co., Sioux Falls, S.D., to Ole Trooien, Hendricks, Minn., enclosed in a letter from Trooien to D.B. Gurney, Feb. 4, 1938.

36. R.A. Rae, Worthing, S.D., to D.B. Gurney, Feb. 3, 1938.

37. D.B. Gurney to R.A. Rae of Worthing, S.D., Feb. 9, 1938.

38. "Form B." NFPTRA Collection.

39. D.B. Gurney to R.A Rae, Worthing, S.D., Feb. 9, 1938.

40. D.B. Gurney to Paul Kantor, Luzerne, Ia., March 8, 1938.

41. Memorandum for Mr. Fres. W. Wallace, Chairman, Nebraska Agricultural Conservation Committee, which quotes from the statement by Martin G. White, enclosed in a letter to D.B. Gurney from Jno. Mueller, March 26, 1938.

42. D.B. Gurney to Jno. Mueller, Clearwater, Neb., March 31, 1938.

43. Saloutos, in *The American Farmer and The New Deal* (Ames: Iowa State University Press, 1982, p. 74), states that county agents tended to see things from the perspectives of the farmers. However, letters received by Gurney suggest that by 1938 at least some American farmers viewed the county agents as unfriendly.

44. Jno. Mueller to D.B. Gurney, March 26, 1938.

45. Hubert Pool, Delano, Minn., to D.B. Gurney, Feb. 13, 1938.

46. Paul Klitzke, Elkton, S.D., to D.B. Gurney, Feb. 15, 1938.

47. William H. Schaller, Groton, S.D., to D.B. Gurney, May 29, 1938.

48. C.E. Pickett, Cozad, Neb., to D.B. Gurney, Feb. 21, 1938,

49. C.H. Compton, Cambridge, Neb., to D.B. Gurney, April 27, 1938.

50. C. p. Knapp of Westfield, Ia., to D.B. Gurney, June 17, 1938.

51. M.A. Brady, Hay Springs, Neb., to D.B. Gurney, December 31, 1938.

52. C.H.T. Jenson, Nelson, Minn., to D.B. Gurney, Feb. 10, 1938.

53. Howard Abbott, Webster City, Ia., to D.B. Gurney, April 6, 1938.

54. Elmer Rohm, Neligh, Neb., to D.B. Gurney, Feb. 14, 1938.

55. F.J. Matezeek, David City, Neb., to D.B. Gurney, Jan. 29, 1938.

56. D.B. Gurney to Carl F. Wall, Mellette, S.D., May 7, 1938.

57. D.B. Gurney to Ed Holscher, Waubay, S.D, May 24, 1938.

58. Gregor Baune, Wabasso, Minn., to D.B. Gurney, February 5, 1938.

59. D.B. Gurney to Albert J. Schmidt, Denison, Ia., June 23, 1938.

60. D.B. Gurney to Kreger Brothers, Bloomfield, Neb., July 2, 1938.

61. "Says New Deal Causes Doubts," *Yankton Press*, April 25, 1938, p. 2.

62. "Entire State GOP Ticket Elected, Republicans Score Over Nation," *Yankton Press*, Nov. 9, 1938, p. 1.

63. D.B. Gurney to Fred Burrs, Dodge Center, Minn., April 2, 1938.

64. NFPTRA Collection. This was 800 paid-up subscribers of the approximately 3,500 to 4,000 farmers who contacted Gurney for information on the program.

65. Jan dePagter, "Gurneys of the Midwest," p. 215. Unpublished manuscript in possession of Gurney family, Yankton, S.D.

66. Edward C. Blackorby, *Prairie Rebel, the Public Life of William Lemke*, Lincoln: University of Nebraska Press, 1963, p. 200.

67. D.B. Gurney to Ben Hurley, Lansing, Ia., March 10, 1938.

68. Donald Van Vleet to John Erp, Feb. 16, 1938, NFPTRA Records.

69. D.B. Gurney to Niel Woltjer, Pennock, Minn., December 22, 1938. NFPTRA Collection.

70. Interview with Charles Ramon Gurney.

71. A.J. Johnson to Joseph Bohohoj, Dodge, Neb., May 26, 1941.

72. Kennedy, *The Fed and the Farmer*, pp. 105, 106.

73. U.S. Congress, Senate Committee on Agriculture, Refund of Processing Tax on Hogs, Hearing, 75th Congress, 3rd Session, March 18, 1938, p. 35.

74. Herbert E. Gaston to H. P. Fulmer, July 25, 1941; Robert R. Chipperfield to George Snively, February 3, 1942. NFPTRA Records.

75. *Ibid.*

76. U.S. Congress, House of Representatives, *Certain Tax Refunds*, Hearings before a Special Subcommittee of the Committee on Agriculture on S2601 and other measures, 75th Congress, 2nd Session, 1938, pp. 2, 3.

77. U.S. Congress, United States Senate, *Refund of the Processing Tax*, Hearings before a Subcommittee of the Committee on Agriculture and Forestry, S.J. Res. 39, 77th Congress 1st Session, May 6, 7, 12, 14, 1938, p. 59.

Chapter 7

1. Richard Kirkendall, *Social Scientists and Farm Politics in the Age of Roosevelt*, University of Missouri Press, 1966, p. 42.

2. Edward L. and Frederich H. Schapsmeier, *Henry A. Wallace of Iowa: The Agrarian Years, 1910–1940*, Ames: Iowa State University Press, 1968, pp. 160–161.

3. *Ibid.*, p. 169.

4. *Ibid.*, pp. 169–170.

5. *Ibid.*

6. John A. Simpson, *The Militant Voice of Agriculture*, Oklahoma City, Oklahoma: Mrs. John A. Simpson Co., 1934, p. 73.

7. Fite, "Farmer Opinion and the Agricultural Adjustment Act, 1933," p. 660.

8. *Ibid.*, p. 661.

9. G.K. Talley to Henry A. Wallace, April 3, 1933, as cited in Fite, "Farmer Opinion and the Agricultural Adjustment Act," p. 661.

10. A.J. Kinnersley to Roosevelt, April 18, 1933. Fite, "Farmer Opinion and the Agricultural Adjustment Act," p. 662.

11. John Horswell to Wallace, April 22, 1933. Fite, "Farmer Opinion and the Agricultural Adjustment Act," p 662.

12. F. Schultheiss to Roosevelt, March 23, 1933. Fite, "Farmer Opinion and the Agricultural Adjustment Act," p. 662.

13. Fred Kuchenmeister to Roosevelt, April 3, 1933. Fite, "Farmer Opinion and the Agricultural Adjustment Act," p. 662.

14. Fite, "Farmer Opinion and the Agricultural Adjustment Act," p. 663.

15. Kennedy, *The Fed and the Farmer*, pp. 72, 73.

16. Senate Committee on Agriculture and Forestry, *Agricultural Emergency Act to Increase Purchasing Power*, Hearings on H.R. 3835, 73rd Congress, 1st session, p. 106.

17. William Z. Ralph to Roosevelt, March 29, 1933. Fite, "Farmer Opinion and the Agricultural Adjustment Act," p. 666.

18. Schapsmeier and Schapsmeier, *Henry A. Wallace of Iowa*, pp. 170–171.

19. *Ibid.*, p. 171.

20. Kirkendall, *Social Scientists and Farm Politics in the Age of Roosevelt*, p. 57.

21. Theodore Saloutos, *The American Farmer and the New Deal*, Ames: Iowa State University, 1982, p. 55.

22. *Ibid.*, p. 9.

23. Murray R. Benedict and Oscar C. Stine, *The Agricultural Commodity Pro-*

grams, Two Decades of Experience, New York: The Twentieth Century Fund, 1956, p. 188.

24. *Ibid.*, p. 196.

25. Lynn W. Eley, "The Agricultural Adjustment Administration and the Corn Program in Iowa, 1933–1940: A Study in Public Administration" (Ph.D. dissertation, University of Iowa, 1952, p. 3.

26. U.S. Department of Agriculture, *Achieving a Balanced Agriculture*, U.S. Government Printing Office, 1934, p. 35.

27. Kennedy, *The Fed and the Farmer*, p. 100; Preliminary Information Blank, "In Account with D.B. Gurney, Yankton, S.D., in Connection with the Matter of Processing Tax Refund." National Farmers Process Tax Recovery Association (NFPTRA) Records, Library Special Collections, Iowa State University.

28. Benedict and Stine, *The Agricultural Commodity Programs*, pp. 187–189.

29. Kennedy, *The Fed and the Farmer*, p. 77.

30. *Ibid.*, p. 23.

31. Simpson, *The Militant Voice of Agriculture*, p. 55.

32. William Hirth to James A. Farley, March 23, 1933. Fite, "Farmer Opinion and the Agricultural Adjustment Act," p. 666.

33. Frank King, King Pig Company, to Franklin Roosevelt, August 7, 1933; S.W. Lund to Frank King, September 16, 1933, NARG 145, AAA 1933–1935, Subject Correspondence, Hogs.

34. "AAA Agents Destroy Pigs While People Starve," *National Union Farmer*, June 15, 1935, p. 1.

35. Milo Perkins to Hollings Randolph, July 5, 1935, NARG 16, Secy's Office, 1935, Hog Cholera-Home Econom. Production Control.

36. Rex. Tugwell to Franklin Roosevelt, August 24, 1935, NARG 16, Secy's Office, Correspondence 1935, Hog Cholera, Hogs.

37. "Wallace Answers Those Who Weep for Little Pigs Killed by Farm Officials," *Daily Republic*, December 17, 1935, p. 4.

38. Milo Reno, "A Glimpse of the Past," Radio Transcript, September 16, 1934, in Roland A. White, *Milo Reno, Farmers Union Pioneer*, New York: Arno Press, 1975, p. 181.

39. H.S. Morgan to Homer Parker, Dec. 2, 1933; Homer Parker to George N. Peek, Dec. 4, 1933; A.G. Black to Homer Parker, Dec. 8, 1933, NARG 145, AAA 1933–1935, Subject Correspondence, Hogs.

40. J.E. Burroughs & Son to M.J. Hart, January 8, 1934; M.J. Hart to J.E. Burroughs, January 13, 1934; M. J. Hart to Henry A. Wallace, January 13, 1934; Henry A. Wallace to M.J. Hart, January 25, 1934, NARG 145, AAA 1933–1935, Subject Correspondence, Hogs, P. T.

41. "Farmers and Hog Producers of DeKalb County, Missouri" to Henry A. Wallace, December 20, 1933, NARG 145, AAA 1933–1935, Subject Correspondence, Hogs.

42. Cecil A. Johnson to W.G. Bishop, August 2, 1934, NARG 145, AAA 1933–1935, Subject Correspondence, Hogs, P. T.

43. M.L. Wilson to Henrik Shipstead, August 7, 1934, NARG 145, AAA 1933–1935, Subject Correspondence, Hogs, P. T.

44. Claude R. Wickard to Emil Priebe, Secretary, Milwaukee Retail Meat Dealers Ass'n, NARG 145, AAA 1933–1935, Subject Correspondence, Hogs, p. T; Mordecai Ezekiel, "Memorandum for the Secretary," August 14, 1935, NARG 16, Secy's Office, Correspondence, 1935, Hog Cholera-Home Ec.

45. Louis H. Bean, Memorandum to Mr. Blaisdell, Jr., January 10, 1934, NARG 145, 1933–1935, Subject Correspondence, Hogs, P. T.

46. Alfred D. Stedman to Chester C. Davis, December 28, 1933, NARG 145, AAA 1933–1935, Subject Correspondence, Hogs, P. T.

47. Ray Anderson to Charles Ebert, January 23, 1935. Reno Papers. Anderson writes that Wallace asked him, but he was unable to tell Wallace where Reno got his funding. Anderson said he "assumed from his question that he believed some interested agency other than a farm group was paying the shot."

48. U.S. Department of Agriculture, Agriculture Adjustment Administration, *Agricultural Adjustment, 1933 to 1935*, Washington, D.C.: Government Printing Office, 1936, p. 176.

49. U.S. Department of Agriculture, *Corn-Hog Adjustment*, Washington, D.C.: Government Printing Office, 1935, p. 13.

50. Wallace Short, "The Corn-Hog Referendum Vote," *Unionist and Public Forum*, October 17, 1935, pp. 1–2.

51. "Corruption in Corn-Hog Vote," *National Union Farmer*, November 15, 1935, pp. 1–3.

52. *Ibid.*

53. Lawrence Oakley Cheever, *The House of Morrell*, Cedar Rapids, Iowa: Torch Press, 1948, p. 223.

54. John S. Wilson to J. p. Hannon, Dec. 20, 1935, NARG 145, AAA 1933–1935, Subject Correspondence, Hogs, P. T.

55. *U.S. v. Butler*, 287 U.S. (1936), in *American Law Reports*, Annotated, pp. 926–938.

56. "F.R. Urges AAA Contract Pay," *Minneapolis Tribune*, January 8, 1936, pp. 1, 11.

57. "Supreme Court Finds AAA Unconstitutional; 6–3 Verdict Dooms Other New Deal Laws; Roosevelt Studies Upset; More Taxes Needed." *New York Times*, January 6, 1936, pp. 1–11.

58. "Emergency Plan Framed," *New York Times*, January 6, 1936, pp. 1,11.

59. Campbell, *The American Farm Bureau and the New Deal*, p. 106; "Friday Parley to Study AAA Substitutes," *Minneapolis Tribune*, January 9, 1936, p. 1.

60. "Farmers Are Split on End of the AAA," *New York Times*, January 7, 1936, p. 1

61. "Farm Bureau Chief Bitter," *Minneapolis Tribune*, January 7, 1936, p. 7.

62. "Iowa Leaders Comment on AAA downfall," *Minneapolis Tribune*, January 7, 1936, p. 7.

63. "Organized Agriculture to Wait Administration Plan for Relief to Farmers; Conference Starts," *Daily Republic*, January 10, 1936, p. 2.

64. "Farmers Will Picket Washington Until New Program Is Accepted," *Daily Republic*, January 13, 1936, p. 2; "Farm Groups Divided on New Agricultural Plan; Grange Draws Own Bill," *Daily Republic*, January 15, 1936, p. 2.

65. Franklin D. Roosevelt, *The Public Papers and Addresses of Franklin D. Roosevelt*, Vol. V, New York: Random House, 1938, p. 48.

66. "Aims Believed Intact in New Farm Program," *Minneapolis Tribune*, January 14, 1936, p. 1.

67. "Soil Conservation Wizard Is Called for Conference on New Farm Legislation," *Daily Republic*, January 11, 1936, p. 1.

68. "Farm Leaders Quiet on Possible Ways to Solve Farm Crisis," *The Daily Republic*, January 8, 1936, p. 2.

69. Oliver Merton Kile, *The Farm Bureau Through Three Decades*, Baltimore, Maryland: Waverly Press, 1948, p. 228.

70. *Ibid.*, p. 229; Henry A. Wallace Papers (Washington, D.C., Library of Con-

gress Photoduplication Service, Microforms, 1971). Henry A. Wallace's appointment book lists an appointment with Ed O'Neal on January 8, 1936, a day before the conference of farm leaders. No other farm leaders seem to have had appointments with Secretary Wallace during that week or the next.

71. "Soil Conservation Wizard Is Called for Conference on New Farm Legislation," *Daily Republic*, January 11, 1936, p. 1.

72. "Report of the Committee of Thirteen to the Agricultural Conference Held at Washington, D.C. Jan. 10–11, 1936," *National Union Farmer*, February 1, 1936, p. 4.

73. "Farm Program Gets Approval," *Minneapolis Tribune*, January 18, 1936, pp. 1, 9.

74. E.E. Kennedy, "National Secretary's Message," *National Union Farmer*, January 15, 1936, p. 2.

75. "Two-Year Farm Plan of Administration Ready for Action in Houses," *Daily Republic*, January 22, 1936, pp. 1, 3.

76. Saloutos, *The American Farmer and the New Deal*, p. 237.

77. "The Soil Conservation Act," Papers *The National Union Farmer*, March 16, 1936, p. 2. See also Henry A. Wallace Papers, "Henry A. Wallace to Donald R. Murphy," February 5, 1936. In a letter to Donald R. Murphy of the Wallace-Homestead Company, Secretary Wallace commented, "Chester [Davis] is right in his element these days working on legislative problems."

78. "Congress Moves to Retain All of Processing Tax Collected," *Daily Republic*, January 14, 1936, p. 1.

79. "Roosevelt and Congressional Heads Agree on Farm Plan; Bankhead Will Introduce Bill," *Daily Republic*, January 21, 1936, pp. 1, 3.

80. Louis H. Bean, "Memorandum to the Secretary," January 23, 1936, NARG 16, Secy's Office, Correspondence, 1936, Taxes.

81. Henry A. Wallace to Franklin Roosevelt, NARG 16, Secy's Office, Correspondence, 1936, Taxes.

82. "Taxes Asked by President," *New York Times*, March 4, 1936, p. 1; "Orders Return of Process Tax," *Minneapolis Tribune*, January 17, 1936, p. 19.

83. "Main Points in Tax Bill," *New York Times*, April 22, 1936, p. 1.

84. Richard Lowitt, editor, *Journal of a Tamed Bureaucrat, Nils A. Olsen and the BAE, 1925–1935*, Ames: Iowa State University Press, 1980, pp. 192, 218.

85. Bureau of Agricultural Economics, *An Analysis of the Effects of the Processing Taxes Levied Under the Agricultural Adjustment Act*, Washington, D.C.: U.S. Treasury Department, Government Printing Office, 1937, pp. 21–36.

86. *Ibid.*, p. 6.

87. *Ibid.*, p. 12.

88. *Ibid.*, pp. 14–15.

89. *Ibid.*, p. 17.

90. *Ibid.*, p. 18.

91. Geoffrey Shepherd, "The Incidence of the AAA Processing Tax on Hogs," *Journal of Farm Economics*, Vol XVII, No. 2, May 1935, pp. 321–339. Another study that also corroborates the studies by the Bureau of Agricultural Economics and Geoffrey Shepherd is the study made by Edwin G. Nourse, Joseph S. Davis and John D. Black, *Three Years of the Agricultural Adjustment Administration*, Washington, D.C.: The Brookings Institution, 1937, pp. 303–304, 437.

92. Bureau of Agricultural Economics, *An Analysis of the Effects of the Processing Taxes Levied Under the Agricultural Adjustment Act*, p. 18.

93. H.W. Christiansen to D.B. Gurney, Feb. 7, 1938; J.J. Wondra to D.B. Gurney,

Feb. 17, 1938; C.C. Ogborn to Cal Polsen, Feb. 10, 1938; D.B. Gurney to H.J. Miller, March 7, 1938; Mrs. J.O. Johnson, March 11, 1938; Tom Murphy to D.B. Gurney, March 30, 1938; D.B. Gurney to H.J. Miller, March 7, 1938. NFPTRA Records.

94. D.B. Gurney to Mrs. J.O. Johnson, March 11, 1938. NFPTRA Records.

95. D.B. Gurney to Mr. Vincent Hackenmueller, March 31, 1938. NFPTRA Records.

96. A.R. Lanies to D.B. Gurney, Feb. 20, 1938. NFPTRA Records.

97. Frank Washeckek to D.B. Gurney, Feb. 9, 1938. NFPTRA Records.

98. A.C. Ewert to D.B. Gurney, April 13, 1938. NFPTRA Records.

99. Dennis O'Neil to D.B. Gurney, June 5, 1938. NFPTRA Records.

100. Frank Lukes to D.B. Gurney, Jan. 31, 1938. NFPTRA Records.

101. Carroll L. Hess to D.B. Gurney, January 9, 1939. NFPTRA Records.

102. Mrs. Geo. Weimer to D.B. Gurney, June 21, 1941. NFPTRA Records.

103. Joseph Filan to D.B. Gurney, April 11, 1938. NFPTRA Records.

104. A.J. Johnson to D.B. Gurney, May 3, 1938. NFPTRA Records.

105. A.J. Johnson to Elmer Paul, August 16, 1939. NFPTRA Records.

106. By 1940 over 50 percent of the farmers in the North Central States were using hybrid corn seed. Yields per acre of corn also increased. They rose from an average of 33 bushels per acre in the early 1930s to 40 or more bushels per acre in the 1940s. Richard J. Schrimper, *Minnesota Corn, Production and Marketing*, St. Paul, Minn.: State-Federal Crop and Livestock Reporting Service, Minnesota Department of Agriculture and Food, 1958, pp. 6, 7.

Chapter 8

1. Edward E. Kennedy, *The Fed and the Farmer*, pp. 40–41.

2. *Ibid.*, p. 49.

3. Jane Wolf Hufft and Anne Nevins Loftis, "Reports of a Downstate Independent, Excerpts from the Letters of Lewis Omer to Allan Nevins, 1930–1953," *Illinois Historical Journal*, Spring 1988, p. 28.

4. "Hancock County's Farm Union Halts Foreclosure Sale," *United Farmers of Illinois*, November 11, 1933, p. 1.

5. John L. Shover, *Cornbelt Rebellion: The Farmers Holiday Movement*, Urbana: University of Illinois Press, 1965, p. 94.

6. "Effective Work Being Done in Kankakee Co.," *The U.F.I.* (published by the United Farmers of Illinois), June 7, 1933, p. 1; "Hancock County's Farm Union Halts Foreclosure Sale," *The U.F.I.*, November 11, 1933, p. 1.

7. "Kankakee and Surrounding Counties on Strike," *United Farmers of Illinois*, November 11, l933, p. 1.

8. Pease, *The Story of Illinois*, p. 244.

9. Jane Wolf Hufft and Anne Nevins Loftis, "Reports of a Downstate Independent," p. 28.

10. Mrs. Margaret Loeffler to Franklin D. Roosevelt, August 15, 1933, NARG 145, 1933–1935, Subject Correspondence, Hogs 1933.

11. Christina McFayden Campbell, *The Farm Bureau, A Study of the Making of National Farm Policy, 1933–1940*, Urbana: University of Illinois Press, 1962. Campbell states, "In Illinois, where the closest of legal ties [between the county agent and the Farm Bureau] was maintained, the farm bureau had the upper hand, at least on the county level" (p. 7). See Theodore Saloutos, *The American Farmer and the New Deal*, Ames: Iowa State University Press, 1982, pp. 47–49, for a discussion on the decision to make county agents the administrators of the local AAA programs in 1933. Also, Gladys Baker, *Cen-*

tury of Service, the First 100 Years of the United States Department of Agriculture, Washington, 1963, p. 160. Baker states that county agents frequently served as secretaries of the corn/hog committees, and sometimes as secretary/treasurers.

12. "Farmers Reject A.A.A. Corn-Hog Plan," *National Union Farmer,* March 15, 1935, pp. 1–2.

13. Edward E. Kennedy, "State of Illinois." National Farmers Process Tax Recovery Association (NFPTRA) Records, Special Collections, Parks Library, Iowa State University.

14. "Radio Address by Edward E. Kennedy … Over the NBC Chain," June 23, 1934. Farmers Union Collection, Historical Collections, University Libraries, University of Colorado at Boulder, pp. 1–2.

15. "AAA Pays Metropolitan Life Insurance Corn-Hog Tax," *National Union Farmer,* May 1, 1935, p. 1.

16. "F.D.R. Silent on Farm Problems in Fireside Chat," *National Union Farmer,* May 1, 1935, p. 1.

17. "Long Bombasts Administration at Convention," *National Union Farmer,* May 1, 1935, p. 1.

18. Alan Brinkley, *Voices of Protest, Huey Long, Father Coughlin, and the Great Depression,* New York: Alfred A. Knopf, 1982, p. 237.

19. Charles J. Tull, *Father Coughlin and the New Deal,* Syracuse: Syracuse University Press, 1965, pp. 93–94.

20. "Father Coughlin Outlines N.U.S.J. Program at Huge Meeting in Detroit, Mich.," *National Union Farmer,* May 1, 1935, p. 1.

21. Kennedy, *The Fed and the Farmer,* pp. 88–93.

22. "Legislative Report," *National Union Farmer,* December 2, 1935, p. 3.

23. "Radio Address by Edward E. Kennedy, National Secretary of the Farmers Union Over National Broadcasting Company May 8, 1936, the Frazier-Lemke refinancing Bill." NFPTRA Records.

24. Alan Brinkley, *Voices of Protest,* Vintage Books: New York, 1983, pp. 133–134.

25. Edward E. Kennedy, Interview June 10, 1989, transcript in Special Collections, Library, Iowa State University, p. 4.

26. Brinkley, *Voices of Protest,* p. 254; Edward C. Blackorby, *Prairie Rebel, the Public Life of William Lemke,* Lincoln: University of Nebraska Press, 1963, pp. 214–215.

27. C.E. Wheaton to Wm. Lemke, January 26, 1936. Lemke Collection.

28. Mr. and Mrs. John Prechtl to William Lemke, May 14, 1936. Lemke Papers.

29. *Ibid.*

30. E.E. Kennedy, Interview, p. 4.

31. "Minutes of the 1936 National Conference of the Union Party," December 19, 1936. Lemke Papers.

32. Farmers Educational and Co-Operative State Union of Nebraska, "Constitution and By-Laws," 1917. Farmers Union Papers, University Libraries, University of Colorado.

33. Resolution of the Washington-Idaho Division, July 8, 1935. Farmers Union Collection, Historical Collections, University Libraries, University of Colorado at Boulder.

34. Blackorby, *Prairie Rebel,* p. 229.

35. E.E. Kennedy to William Lemke, November 24, 1936; William Lemke to E.E. Kennedy, November 24, 1936; William Lemke to E.E. Kennedy, November 30, 1936; William Lemke to E.E. Kennedy, December 3, 1936; E.E. Kennedy to William Lemke, December 5, 1936. Lemke Papers.

36. *Kennedy's Washington Letter,* No. 1 April 5, 1937. Lemke Papers.

37. Edward E. Kennedy to A.J. Johnson, June 23, 1937. NFPTRA Records.

38. "Agreement" between the National Farmers Process Tax Recovery Association and Fred Wolf, September 17, 1935. NFPTRA Records.

39. Illinois Counties Claims and Payments. NFPTRA Records.

40. Donald Van Vleet to Fred Wolf, April 16, 1938. NFPTRA Records.

41. C.F. Schulz to National Farmers Process Tax Recovery Association, December 8, 1937. NFPTRA Records.

42. Donald Van Vleet to C.F. Schulz, December 14, 1937. NFPTRA Records.

43. Dr. Glenn Smith to A.J. Johnson, September 7, 1938. NFPTRA Records.

44. A.J. Johnson to Alfred Harm, January 5, 1939. NFPTRA records.

45. A.J. Johnson to John p. Lingenfelter, April 13, 1939. NFPTRA Records.

46. E.E. Kennedy to A.J. Johnson, July 13, 1938. NFPTRA Records.

47. Lowell K. Dyson, *Red Harvest, the Communist Party and American Farmers*, Lincoln: University of Nebraska Press, 1982, p. 148; "The Minnesota and Michigan Case, Charter Revoked by National Convention." NFPTRA Papers.

48. Fred Winterroth to A.J. Johnson, May 6, 1938. NFPTRA Records.

49. Fred Winterroth to A.J. Johnson, June 16, 1938. NFPTRA Records.

50. Robert Spencer to John Vesecky, June 25, 1938. Kennedy Collection, Univ. of Iowa.

51. Robert Spencer to John Vesecky, June 25, 1938. Kennedy Papers, University of Iowa.

52. A.J. Johnson to Fred Winterroth, June 9, 1938; Edward E. Kennedy to A.J. Johnson, August 25, 1938. NFPTRA Records.

53. Dyson, *Red Harvest*, p. 148.

54. Paul G. Erickson to Board Members of National Farmers Union, Feb. 10, 1939. Farmers Union Papers.

55. L.S. Ahrendt to James G. Patton, January 12, 1939. James Patton Collection.

56. Robert Spencer, "Call of Committee on New Farm Organization," January 16, 1939. NFPTRA Papers.

57. Letterhead of the 1941 National Farmers Guild. For more information on the National Farmers Guild, see also Lowell Dyson's *Farmers' Organizations*, New York: Greenwood Press, 1986, pp. 202–204.

58. "Resolutions adopted by the National Farmers Guild Convention," Mansfield, Ohio, November 1941. Kennedy Papers.

59. Robert Spencer, "Call of Committee on New Farm Organization," Jan. 16, 1939. Kennedy Papers.

60. Dyson, *Red Harvest*, p. 148; "Patton Is Willing," *Time*, September 14, 1942, p. 22.

61. "New Tax Misleading to Farmers," *Social Justice*, March 27, 1936, p. 12.

62. "Who Signed the Frazier-Lemke Petition," *Social Justice*, March 13, 1936, p. 3.

63. "How Mr. Hull Stabs the Farmers," *Social Justice*, October 23, 1939, p. 12.

64. "Cost of Production; the Key to Lasting Solution of Farm Problem," *Social Justice*, December, 18, 1939, p. 5.

65. "Farmers Tired of New Deal Quack Measures," *Social Justice*, January 1, 1940, p. 15.

66. "Wallace Report Sees Continued Farm Control," *Social Justice*, January 22, 1940, p. 5.

67. "Farmers Sour on New Deal Control of Crops," *Social Justice*, July 24, 1939, p. 9.

68. "New Deal Policies Bring Income Drop to Farmers," *Social Justice*, June 19, 1939, p. 12.

69. "E.C. Kennedy Outlines Plan," *Social Justice*, March 12, 1936, p. 8.

70. Joseph P. Wright, "The Farmer Goose-Steps," *Social Justice*, March 7, 1938, p. 5.

71. S. Fred Cummings, "The Farmer Suffers Most," *Social Justice*, January 30, 1939, p. 16.

72. "Farm Revolt Wins," *Social Justice*, August 29, 1938, p. 6.

73. "These Farmers Have the Right Slant on Things," *Social Justice*, November 6, 1939, p. 13.

74. Roy Wortman, "Coughlin and the NFU," unpublished paper, 1993.

75. F.R. Lennox to Edward E. Kennedy, 1936. Ohio Farmers Union records, Ohio Farmers Union, Ottawa, Ohio. Quoted in Roy Wortman, "Coughlin in the Countryside," unpublished manuscript, p. 21.

76. George Edward Sullivan, "America's Insidious Foes," *Social Justice*, December 5, 1938, pp. 10–11.

77. *Ibid.*

78. "Powers Plan Aid to Jewish Refugees," *Social Justice*, July 25, 1938, p. 20.

79. "Salazar of Portugal Builds Bulwark Against Paganism," *Social Justice*, June 12, 1939, p. 1

80. J.S. Barnes, "The War Profiteers," *Social Justice*, December 11, 1939, p. 11.

81. J.S. Barnes, "Italy's New School Charter," *Social Justice*, July 31, 1939.

82. Ben Marcin, "The Truth About the Protocols," *Social Justice*, September 3, 1938, p. 10.

83. Charles J. Tull, *Father Coughlin and the New Deal*, Syracuse, N.Y.: Syracuse University Press, 1965, pp. 93–94.

84. Marcin, "The Truth About the Protocols," p. 11.

85. Harry Lux interview by John Shover. University of Iowa Archives.

86. *Ibid.*

87. John Shover, *The Cornbelt Rebellion*, pp. 182–183.

88. *Ibid.*

89. Lowell Dyson, *Farm Organizations*, pp. 201–202; Carl Mote, *The New Deal Goose Step*, New York: Daniel Ryerson, Inc., 1939, pp. 95–123.

Chapter 9

1. Roger Babson, "Bumper Crops for the U.S.," *Peoria Journal*, May 8, 1938, p. 2.

2. "Farmers Howl Down Backer of Control Law," April 18, 1938. From Ruebush-Goodpasture Collection, Archives, Western Illinois University; "Farmers Howl Down Backer of Control Law," *Chicago Tribune*, April 19, 1938, p. 1.

3. "The Farmers' Awakening," *Peoria Journal*, April 21, 1938, p. 10; "Revolt Against Control Sweeps Over Corn Belt," *Chicago Tribune*, April 20, 1938, p. 5.

4. "Farmers! Urge Your Friends and Acquaintances...." Ruebush-Goodpasture Collection.

5. "Farm Revolt Spreads Over Corn Control," April 27, 1938. Ruebush-Goodpasture Collection.

6. "3,500 Farmers from Four of Corn Belt States Pledge Opposition to Federal Crop Control," *New York Times*, April 28, 1938, p. 2.

7. "Stage Is Set for Meeting of Farmers," *Daily Journal*, n.d.

8. "Consider Change in Crop Control Program After Macomb Protest Meeting," *Daily Journal*, April 28, 1928, pp. 1–2.

9. "Armory Jammed to Overflowing: Many States, Counties Represented," *Daily Journal*, April 18, 1938, p. 1.

10. "Objectors to Farm Program Cheer Speakers," *Daily Journal*, April 28, 1938, p. 2.

11. "Farmers Rush to Join War on Crop Control," *Chicago Herald and Examiner*, April 28, 1938, pp. 1–2.

12. "Consider Change in Crop Control Program After Macomb Protest Meeting," *Daily Journal*, April 28, 1938, p. 1.

13. "Armory Jammed…," p. 1.

14. Mrs. Frank Taylor, "Views of Others," *Daily Journal*, April 21, 1928.

15. Mrs. Tom Joy, "Views of Others," *Daily Journal*, n.d.

16. Mrs. W.L. Heberer, "Vote on Quotas," *Daily Journal*, n.d.

17. "Corn Belt League Plans Many Meetings," May 5, 1938. Ruebush-Goodpasture Collection.

18. "500 Warren County Farmers Join Revolt on Crop Control." Ruebush-Goodpasture Collection; "Farmers Hoot Wallace Aid on Crop Control," *Chicago Tribune*, May 6, 1938, p. 1.

19. Homer Neisz, "500 Farmers Vote to Oppose U.S. Crop Rule," *Chicago Daily Tribune*, May 7, 1938.

20. "Local Farmers Form Liberty League; Dawson, President; Peterson, Secretary," *Bureau County Republican*, May 12, 1938, pp. 1, 6.

21. *Ibid.*; "Speakers Lampoon Agricultural Act," *Bureau County Republican*, May 2, 1938, pp. 1, 2.

22. Paul D. Shoemaker, "Future of the Farm 'Revolt' Is Considered," *Daily Journal*, n.d. Ruebush-Goodpasture Collection.

23. Morse to Burg, April 25, 1938. Foster Papers.

24. "Announcement, Headquarters of the Corn Belt Liberty League," n.d. Ruebush-Goodpasture Collection.

25. "Local Farmers Form Liberty League; Dawson, President; Peterson, Secretary," *Bureau County Republican*, May 12, 1938.

26. Hulda Ruebush Interview conducted by Lynnita Summer, March 31, 1990.

27. *Ibid.*

28. George Thiem, "Finley Foster Not on Relief, He Tells World." Ruebush-Goodpasture Collection.

29. Interview with Mrs. Carl Ruebush.

30. "Corn Leaguers Plan Meetings in 4 Counties." Ruebush-Goodpasture Collection.

31. "No Overproduction in U.S., Says F. Dick," n.d. Ruebush-Goodpasture Collection.

32. Willis Thornton, *Peoria Star*, June 7, 1938, p. 10.

33. "Purposes for Which the Corn Belt Liberty League Is Incorporated," Resolutions of Stary County Corn Belt Liberty League, June 10, 1938. Ruebush-Goodpasture Collection.

34. "Liberty Leaguers of the Soil Want to Manage Their Own Farms, Corn Belt 'Revolt' Is Against AAA," *American Liberty Magazine*, August 4, 1938, p. 4. Ruebush-Goodpasture Collection.

35. Lynnita Aldridge Sommer, "Illinois Farmers in Revolt, the Corn Belt Liberty League," *Illinois Journal of History*, p. 231.

36. Charles N. Wheeler, "Farmers United to Fight Crop Cut in Indiana," April 28, 1938; "Farm Revolt on Corn Gains Over Indiana," April 29, 1938. Ruebush-Goodpasture Collection.

37. "Corn Belt Liberty League Holds Large Meeting in Indiana," *American Liberty Magazine*, August 4, 1938, p. 12.

38. "Report Good Meeting of League in Indiana," n.d. Ruebush-Goodpasture Collection.

39. Carlyle Hodgkin, "Teft Tells About Liberty League," *Nebraska Farmer*, June 4, 1938, p. 9.

40. "Washington Hears from Farmers," *The Nebraska Farmer*, May 21, 1938, p. 7.

41. Stanley Morse to Dan Casement, April 28, 1938. Dan D. Casement Papers, University Archives, Kansas State University.

42. Thale Skovgard to Dan Casement, April 27, 1938. Dan D. Casement Papers.

43. V.M. Reed to Dan Casement, May 2, 1938. Dan D. Casement Papers.

44. C.L. Potter to Dan Casement, May 6, 1938. Dan D. Casement Papers.

45. H.E. Gordon to Dan Casement, May 11, 1938. Dan D. Casement Papers.

46. Ballard Dunn to Casement, May 6, 1936. Dan D. Casement Papers.

47. George Herzog to Dan Casement, May 11, 1938. Dan D. Casement Papers.

48. Tilden Burg to A.J. Ostlund, May 11, 1936. Dan D. Casement Papers.

49. Ed Harvey to Dan Casement, May 14, 1938. Dan D. Casement Papers.

50. Stanley Morse to Dan Casement, May 14, 1938. Dan D. Casement Papers.

51. H.E. Gordon to Dan Casement, May 11, 1938. Dan D. Casement Papers.

52. J.J. Lilley to Dan Casement, May 27, 1938; George A. Hunt to Dan Casement, May 28, 1938. Dan D. Casement Papers.

53. A.C. Ostlund to Dan Casement, May 30, 1938; Charles Kishop to A.J. Ostlund, May 27, 1938. Dan D. Casement Papers.

54. Stanley Morse to Dan Casement, May 31, 1938. Dan D. Casement Papers.

55. Stanley Morse to Dan Casement, June 3, 1938. Dan D. Casement Papers.

56. S.M. Swenson to Dan Casement, June 30, 1938. Dan D. Casement Papers.

57. Dan Casement, Thale p. Skovgard and E.C. Davis, "Letter Sent to Area Farmers," July 1938. Dan D. Casement Papers.

58. S.M. Swenson to Dan Casement, July 1, 1938. Dan D. Casement Papers.

59. Helen Lobdell to Dan Casement, July 7, 1938; V.E. Becannon to Dan Casement, July 18, 1938; V.E. Hawkinson to Dan Casement, July 26, 1938; Dan D. Casement Papers.

60. H.E. Gordon to Dan Casement, August 3, 1938. Dan D. Casement Papers.

61. Stanley Morse to Dan Casement, August 8, 1938. Dan D. Casement Papers.

62. *Ibid.*

63. Stanley Morse to Dan Casement, August 9, 1938; Dan Casement to Stanley Morse, n.d. Dan D. Casement Papers.

64. Stanley Morse to Herbert Hoover, August 27, 1938. Herbert Hoover Papers.

65. "Fire on 'Wheat Bribe,'" *Kansas City Star*, August 16, 1938.

66. "Farm Revolt Wins," *Social Justice*, August 29, 1938; p. 6.

67. Edwin Lindstrom to Dan Casement, September 19, 1938; Victor Hawkinson to Dan Casement, September 12, 1938; J.W. Wilson to Dan Casement, September 9, 1938; Conrad Crome to Dan Casement, September 8, 1938. Dan D. Casement Papers.

68. Stanley Morse to Herbert Hoover, October 8, 1938. Herbert Hoover Papers.

69. John M. Collins, "Opposition to AAA Spreads in Kansas," *New York Times*, September 25, 1938, p. 25, Section III.

70. "Wallace Clarifies Aim for AAA as Envisaged in Federal Policy," *New York Times*, October 2, 1938, pp. 1,6, Section III.

71. Joe Cooperstein to Dan Casement, September 30, 1938. Dan D. Casement Papers.

72. Fred Grantham to Dan Casement, September 30, 1938. Dan D. Casement Papers.

73. W.F. Thompson to Dan Casement, September 30, 1938. Dan D. Casement Papers.

74. G.W. Norris to Dan Casement, September 30, 1938. Dan D. Casement Papers.

75. Elias Farr to Dan Casement, October 8, 1938. Dan D. Casement Papers.

76. Thomas F. Doran to Dan Casement, October 10, 1938. Dan D. Casement Papers.

77. C.S. Walker to Dan Casement, October 19, 1938. Dan D. Casement Papers.

78. Fred Ramsey to Dan Casement, October 9, 1938; Newspaper Release, October 20, 1938; Thale Skovgard to Dan Casement, October 17, 1938. Dan D. Casement papers.

79. Stanley Morse to Dan Casement, October 13, 1938. Dan D. Casement Papers.

80. "Farm Income Drop Forecast, Officials Cite Many Adverse Factors," *Peoria Journal*, January 1, 1938, p. 2.

81. Kurt McKelvie, *The Nebraska Farmer*, January 15, 1938, p. 6.

82. "Farmers Howl Down Backer of Control Law," *Chicago Tribune*, April 19, 1938, p. 1.

83. "Farm Revolt Spreads Over Crop Control," April 27, 1938. Ruebush-Goodpasture Collection.

84. "IAA says 35¢ Corn Possible in the Fall," *Daily Journal*, May 2, 1938, p. 10.

85. August H. Andresen, "Farmers Fight to Retain Constitutional Freedom," *Congressional Record*, May 2, 1938.

86. *Ibid.*, p. 1.

87. *Ibid.*, p. 2; "Charges Spies at Local Farm Meeting," May 3, 1938; "AAA Made Target of Spy Charge," May 2, 1938. Ruebush-Goodpasture Collection.

88. "Grumbling, But No Revolt in the Farm Areas," *New York Times*, May 15, 1938, p. 3, Section IV.

89. "AAA Meeting to Fight League Is Washington Plan," *Daily Journal*, April 30, 1938.

90. "Washington Set for Attack on Farm Revolt." Ruebush-Goodpasture Collection.

91. "Critics of AAA Answered by Earl Smith," *Daily Journal*, May 31, 1938, p. 10.

92. "Grumbling, But No Revolt in the Farm Areas."

93. "Wallace Hits AAA Critics," *Peoria Journal*, May 13, 1938, p. 7.

94. Chesley Manley, "Revolt Forces Concessions on Crops Rule Act," *Chicago Tribune*, May 5, 1938, pp. 1, 8. Reporters speculated that perhaps the fact that cotton and tobacco farmers obtained changes in the law, and corn farmers of Illinois did not, was because Southern senators were concerned, but the Illinois senators did not show any particular interest in the program.

95. William Hirth, "The Farm Revolt Grows," *The Missouri Farmer*, June 1, 1938, p. 8.

96. "Are We Governed Too Much?" *Sentinel*, Greenleaf, Kansas. Published in *Daily Journal*, June 13, 1938, p. 4.

97. "Crop Control Assailed," *New York Times*, July 29, 1938, p. 27.

98. "Crop Big but Corn Quotas May Be Avoided," *Daily Journal*, July 21, 1938, p. 1; "No Corn Referendum," *Moline Dispatch*, n.d. Ruebush-Goodpasture Collection.

99. J.E. Poole to Dan Casement, September 7, 1938. Dan D. Casement Papers. "Farmers Divided on Acreage Plans," *New York Times*, August 21, 1938, p. 1, Section III.

100. Founders and Incorporators to Members of the Corn Belt Liberty League, n.d. Ruebush-Goodpasture Collection.

101. E.R. Illenden to Dan Casement, September 8, 1938. Dan D. Casement Papers.

102. Thomas Doran to Dan Casement, August 31, 1938. Dan D. Casement Papers.

103. "Farmers Divided on Acreage Plans."

104. John H. Crider, "Farm Price Slump Tries AAA Program, Wallace Faces Difficult Task of Holding Farmer to Idea in a Campaign Year." *New York Times*, August 14, 1938. p. 6.

105. "'Farm Revolt' Stirs AAA to Plead for Patience," *New York Times*, October 9, 1938, p. 6, Section IV.

106. "Wallace Defends Aims in Corn Belt," *New York Times*, October 15, 1938, p. 6.

107. Lionel D. Eyman, "Fight for 'Best Farm Act,' Asks Wallace," *The American Liberty Magazine*, October 21, 1938, p. 1. Ruebush-Goodpasture Collection.

108. Luther A. Huston, "Vote in Pivotal States to Offer Guides on 1940"; Henry D. Dorris, "G.O. P. Looks for 'Working Minority'," *New York Times*, November 6, 1938, p. 6E.

109. Turner Catledge, "Shifts in Congress"; Arthur Krock, "Win Back 10 States," *New York Times*, November 9, 1938, p. 1.

110. Congressional Quarterly, *Guide to U.S. Elections*, Washington, D.C.: Congressional Quarterly Inc., 1975, pp. 485–509.

111. Congressional Quarterly, *Presidential Elections Since 1789*, Washington, D.C., 1975, pp. 786–790.

112. "G.O. P. Gains Favor C.B.L.L.," *The American Liberty Magazine*, November 19, 1938, p. 1.

113. Kurt Grunwald to Dan Casement, November 9, 1938; Alex Smith to Dan Casement, November 9, 1938. Dan D. Casement Papers.

114. Stanley Morse to Herbert Hoover, November 14, 1938. Hoover Papers.

115. Herbert Hoover to Stanley Morse, November 17, 1938. Hoover Papers.

116. "Predict Change in '39 Farm Law," *The American Liberty Magazine*, December 22, 1938, p. 1.

117. Frank Murphy to James Farley, December 7, 1938. Franklin D. Roosevelt Library.

118. Dan Mason to Franklin D. Roosevelt, November 15, 1938. Franklin D. Roosevelt Library.

Chapter 10

1. John L. Shover, *Cornbelt Rebellion: The Farmers Holiday Movement*, Urbana: University of Illinois Press, 1965, p. 88.

2. Donald Van Vleet to Representative Vincent Harrington, October 13, 1937. NFPTRA Records.

3. Wilfred Were to Wm. Lemke, Feb. 12, 1940. Lemke Papers.

4. Wm. Lemke to Glen T. Stebbins, Feb. 14, 1939; Karl Mundt to Secretary Henry A. Wallace, May 20, 1939; Wm. Lemke to Karl Mundt, May 23, 1939. Lemke Papers.

5. Donald Van Vleet, "State President's Editorial, Getting Acquainted," *Iowa Union Farmer*, October 25, 1941, p. 1.

6. Edward E. Kennedy, *The Fed and the Farmer*, back cover.

7. A.J. Johnson Family, Interview, Moorhead, Iowa, June 7, 1991.

8. Edward C. Blackorby, *The Prairie Rebel*, Lincoln: University of Nebraska Press, 1963, pp. 276-277.

9. Charles Ramon Gurney, Interview, Iowa City, August 2, 1990.

10. Clarke, Sally H. *Regulation and the Revolution in United States Farm Productivity*, Cambridge University Press, 1995, p. 161.